THE
MATHEMATICAL
TOURIST

THE
MATHEMATICAL
TOURIST

New and Updated Snapshots of Modern Mathematics

Ivars Peterson

W. H. Freeman and Company
New York

COVER DESIGN: Patricia McDermond

INTERIOR DESIGN: Victoria Tomaselli

Library of Congress Cataloging-in-Publication Data

Peterson, Ivars.
    The mathematical tourist : new and updated snapshots of modern
mathematics / Ivars Peterson.
        p.   cm.
    Includes bibliographical references and index.
    ISBN 0-7167-3250-5
    1. Mathematics—Popular works.   I. Title.
    QA93.P377   1998
    510—dc21   98-2697
                                                                CIP

Printed in the United States of America

First printing 1998

To our sons,
Eric and Kenneth,
and in memory of
Frederick J. Almgren Jr. (1933–1997)

CONTENTS

Contents

—— 3 ——

Twists of Space   53

—— 4 ——

Shadows from Higher Dimensions   87

Contents

Contents

A COLOR GALLERY OF MODERN MATHEMATICS

PLATE 1   Crystal Tiles
A mosaic made up of fat and thin diamond shapes is one example of a Penrose tiling. This curious mathematical structure has fivefold symmetry, as seen in the starlike figures in the pattern. Decorating each tile with colored bands and imposing the rule that tiles be joined only if bands of the same color run continuously across a boundary ensure that the tiles do not fall into a periodic pattern.

PLATE 2   Decagon Covers
This novel pattern, made up of overlapping 10-sided tiles, can serve as a model of how atoms sometimes arrange themselves into a complex quasicrystalline structure instead of the regularly repeating units of an ordinary crystal.

PLATE 3   Minimal Möbius
Glue together the ends of a long strip of paper after giving one end a half twist. The result is a remarkable figure called a Möbius strip, which has only one side and one edge. The same basic form also lies at the heart of a certain infinite minimal surface that twists around to intersect itself.

PLATE 4   Smooth Moves
Graceful contours sweep out from the core of a recently discovered example of an infinite minimal surface that doesn't intersect itself. This particular surface can be modeled on a thrice-pricked doughnut. Two of the doughnut's puncture points are stretched out to form the hourglass shapes, while the third becomes the figure's equatorial skirt.

PLATE 5   Psychedelic Onion
A sphere isn't the only closed surface that has the same nonzero curvature everywhere. A self-intersecting torus, which doesn't look at all like a sphere, has a similar property. Peeling away the outside of this toroidal onion reveals an intricate inner geometry.

PLATE 6   Sliced Doughnuts
A sphere can be sliced into a sequence of parallel circles and a pair of points at opposite poles. Similarly, a four-dimensional hypersphere can be visualized as two linked circles and a succession of surrounding doughnut-shaped surfaces. Rotating this representation of a hypersphere produces a complex pattern of intertwined rings.

PLATE 7   Mountains of the Mind

The jagged path of a random walk, in which each step up or down is determined by chance, can be thought of as a cross section through a rugged, mountainous landscape. Such cross sections, along with appropriate shading and colors, can be used in computer graphics to construct a surprisingly realistic rendition of a mountain scene.

PLATE 8   Diffusion Tree

One way to grow a tree is to let particles float about randomly until they hit and stick to a previously generated particle cluster. One by one, the particles aggregate into a treelike shape. The computer version of this process is called diffusion-limited aggregation. The colors correspond to the order in which particles arrive at the tree.

PLATE 9   Branching to Infinity

Start with a number. Run it through an equation that generates a new number, then repeat the operation on the answer, and so on. The result is a sequence of numbers that, depending on a control parameter built into the equation, may eventually settle on a single value, cycle through a small group of values, or randomly hop from one value to another. For a given equation, a computer can plot which of these possibilities occur for different parameter values.

PLATE 10   Culture Clash

Magnifying a portion of the Mandelbrot set's boundary reveals a chaotic world of winding tendrils, ragged shapes, and miniature copies of the snowmanlike Mandelbrot figure. Different colors show the rates at which various points near the boundary escape to infinity.

PLATE 11   The Ubiquitous Snowman

Surprisingly similar features show up when both simple and complicated functions are iterated. For example, the pimply plump figure characteristic of the Mandelbrot set often makes surprise appearances in the middle of maps associated with many different functions.

PLATE 12   Newton's Roots

Newton's method is a venerable technique for finding the roots, or zeros, of an equation. The expression $z^4 - 1$, where z is a complex number, has four roots. In the complex plane, each root is surrounded by a basin of attraction. Normally, Newton's method nudges a starting point selected from a particular basin toward the appropriate root. However, points near the convoluted boundaries between basins may behave in unexpectedly chaotic ways.

PLATE 13   Bursting into Chaos

The exponential function, which can be used to represent compounded growth, can sometimes explode into chaos when the value of a parameter reaches a critical value. Such a system may provide a useful model of physical phenomena, such as combustion, that involve rapid transitions between chaotic and stable states.

PLATE 14   Grid Rules

Cellular automata may consist of lines of cells, where each color represents one of five possible states. The value of each cell is determined by a simple rule based on the values of its neighbors on the previous line. The top pattern is grown from a single colored cell, whereas the lower pattern begins as a random mix of colored cells.

PLATE 15   Snowflake Cells

Starting with a single red cell in the center of a lattice of small hexagonal cells, a computer program step by step generates a snowflakelike pattern. The successive layers of ice are shown as a sequence of colors, ranging from red to blue every time the number of layers doubles.

PLATE 16   Ant Tracks

A set of simple rules activates a virtual ant. Wandering across an infinite checkerboard, this cybercritter leaves intriguing patterns in its wake, providing food for thought for mathematicians and others interested in the behavior of cellular automata.

PREFACE

"Curiouser and curiouser," cried Alice.
—Lewis Carroll, *Alice's Adventures in Wonderland*

Years ago, as a University of Toronto undergraduate majoring in physics and chemistry (with a heavy dose of mathematics on the side), there were times when I felt I had dropped down a rabbit hole into a bewildering land. More than once, I remember sitting in cavernous lecture halls, surrounded by dozing or fidgeting classmates, trying to figure out what was going on. The lecturer would be scrawling equation after equation across the blackboard and speaking in a puzzling language that sounded like English but somehow wasn't. Although I could pin a meaning to practically every word, I didn't seem to understand anything he said. I was as lost as an accidental tourist wandering in a very foreign country.

That disconcerting feeling hasn't altogether disappeared. Now, as a journalist writing about science and mathematics, I often have the same dismaying sense of being an outsider when I listen to researchers presenting papers at professional meetings. The words sound familiar, but the language seems so foreign. During the eight years that I taught high school science and mathematics, I suppose my students also experienced the same feeling, despite my best efforts to convey a sense of the fun and excitement to be found in those subjects.

Professional mathematicians, in formal presentations and published papers, rarely display the human side of their work. Frequently missing among the rows of austere symbols and lines of dense type and among the barely legible mathematical formulas marching across transparencies projected on a screen is the idea of what their work is all about—how and where their piece of the mathematical puzzle fits, their fountains of inspiration, and the images that carry them from one discovery to another.

To most outsiders, modern mathematics represents unknown territory. Its borders are protected by dense thickets of technical terms, its landscapes strewn with cryptic equations and inscrutable concepts. Few realize that the world of modern mathematics is rich with vivid images, provocative ideas, and useful notions.

In recent years, the mathematical community as a whole has displayed a renewed emphasis on applications, encouraged a return to concrete images, and allowed an increasing and more explicit role for mathematical experiments. Computation and computer graphics have provided a wide range of colorful, exotic images to illuminate mathematical ideas and suggest new mathematical questions. These changes are making the forbidding territories of mathematics more accessible than before to outsiders.

Gazing into Math Land, one may now catch the glint of a fractal tower piercing a wispy mist or feel the inexorable pull of a swirling strange attractor. Sometimes the air murmurs with snatches of wondrous tales about mathematicians tangling with knots, peering into higher dimensions, pursuing digital prey, playing with soap bubbles, or wandering in labyrinths.

A decade after the publication of the original edition of *The Mathematical Tourist*, I have taken the opportunity to revise my vision of modern mathematics and to introduce readers to several startling discoveries and mathematical advances that have occurred during the intervening years. In particular, this new edition updates efforts aimed at identifying large prime numbers and highlights the recent, astonishing, and quite unexpected finding that the use of computers based on quantum logic significantly speeds up the factoring of composite numbers.

The revised, updated book also includes new material about the structure of crystals and about discoveries illuminating the relationship between four-dimensional geometry and physical theories of the nature of time, space, and matter. It adds the application of cellular automata models to social questions and to the peregrinations of virtual ants. It extends the discussion of applications of fractals and introduces the notion of controlling chaos. A greatly expanded concluding chapter on the nature of mathematical proof includes accounts of Turing machines, transparent proofs, and how a computer program managed to prove a long-standing mathematical conjecture. It ends with the dramatic unveiling of a proof by Andrew Wiles of Fermat's last theorem.

In more than 15 years of reporting for *Science News*, I've had a chance to talk with mathematicians responsible for changing the look and style of modern mathematics. I've seen many wonderful visualizations of deep mathematical concepts. Some, such as the fractal images created by Heinz-Otto Peitgen and his colleagues, even show up in art galleries and tour the world. I've learned that mathematics is full of intriguing, unanswered questions and that new discoveries and modes of thinking keep pushing mathematics onward.

Mathematics is a vast enterprise. Thousands of pages of original research, spread across hundreds of journals, are published every year throughout the world. The United States has nearly 10,000 mathematicians in academia, comparable to the numbers of physicists and chemists. The topics in this book represent only a small fraction of what goes on in the field—just a few snapshots taken during brief forays into the sometimes bizarre and always fascinating wonderland of modern mathematics. This volume is meant as an informal introductory guidebook for travel in the world of contemporary mathematics.

Preface

The scholarly writings of mathematicians are, in general, liberally sprinkled with footnotes and references to antecedents. In this way, mathematicians credit earlier work on which the structure of mathematics is carefully built. Conjectures are often associated with certain individuals; proofs may have a long history of partial successes that fence in conjectures more and more tightly. To save the reader from having to go through long catalogs of names associated with particular ideas, I have chosen to concentrate on certain central concepts. Similarly, I have usually mentioned only the key mathematicians responsible for a given piece of research. Books and articles listed in the Further Reading section fill in many of the details.

Because I'm not a native of the land of mathematics, I've had to rely on a host of mathematicians to guide me through mathematical thickets. I've borrowed from the works of many mathematicians and scientists, gleaning ideas and examples from numerous lectures, research papers, articles, and conversations. Lynn Arthur Steen and Ronald Graham have been particularly helpful in pointing me toward what is new in mathematics. Ian Stewart's articles in *Nature*, *The Mathematical Intelligencer*, and *New Scientist*, along with Barry Cipra's writings for *Science*, *SIAM News*, and the American Mathematical Society, have also proved extremely useful. References that I found especially valuable are noted in the Further Reading section.

I am grateful to my editors at *Science News*, Joel Greenberg, Patrick Young, and Julie Ann Miller, for allowing me to wander in obscure mathematical pastures in search of stories about mathematics and mathematicians. Much of the material in this book has appeared in a somewhat different form in *Science News*.

I wish to thank the following mathematicians and scientists for helping me by explaining ideas, providing illustrations, or supplying reference materials for both the original and revised editions of *The Mathematical Tourist*: Robert Axtell, Laszlo Babai, Tom Banchoff, Michael Barnsley, Harold Benzinger, Joan Birman, Manuel Blum, Ernie Brickell, Scott Burns, John Cahn, Nicholas Cozzarelli, Jim Crutchfield, Bob Devaney, Keith Devlin, A. K. Dewdney, Persi Diaconis, Rick Durrett, Joshua Epstein, Jim Fisher, Howland Fowler, Martin Gardner, Solomon Golomb, Andrew Granville, Branko Grünbaum, John Guckenheimer, Richard Guy, Joel Hass, David Hoffman, James Hoffman, John Hubbard, Leo Kadanoff, Hüseyin Koçak, David Laidlaw, Edward Lorenz, Benoit Mandelbrot, David McQueen, Paul Meakin, Ken Millett, Rick Norwood, Peter Oppenheimer, Norman Packard, Julian Palmore, Heinz-Otto Peitgen, Roger Penrose, Charles Peskin, Carl Pomerance, Jim Propp, Paul Rapp, Peter Renz, Rudy Rucker, Manfred Schroeder, Peter Shor, Gus Simmons, Paul Steinhardt, John Sullivan, Scott Sutherland, Jean Taylor, Bill Thurston, Anthony Tromba, Hugh Williams, Stephen Wolfram, and Larry Wos. My apologies to anyone I have inadvertently failed to include in the list.

Preface

I am grateful to my wife, Nancy, for many helpful suggestions and to the numerous readers of the original edition of *The Mathematical Tourist* for their enthusiastic response to the book. I also greatly appreciate the efforts of everyone at W. H. Freeman and Company who worked so hard—now on two occasions—to transform a thick stack of manuscript pages and rough illustrations into the finished book.

1

Explorations

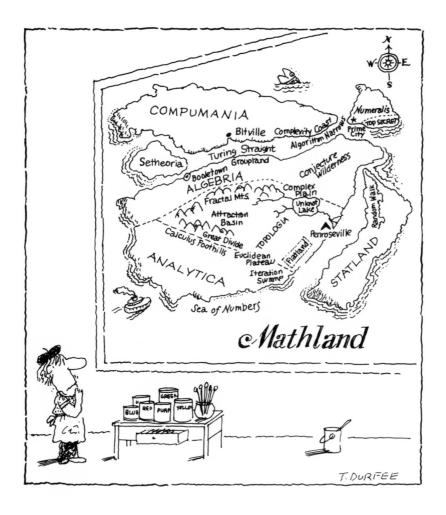

A map is a picture of both the known and the unknown. In olden times, travelers carried maps showing major cities and well-trodden trade routes alongside mysterious territories marked by fanciful names and decorated with mythical creatures. What wasn't known was guessed at. It was these unknown regions, with their promise of adventure and vague hints of fabulous treasure, that attracted early explorers.

A map of modern mathematics would reveal a similar mix of the familiar, the exotic, and the unknown. Algebra, trigonometry, and Euclidean geometry, familiar to high-school students, lie in well-settled areas. Newer settlements, such as calculus and topology, establish their spheres of influence nearby. Upstarts—computer science, for one—gnaw at old boundaries. Beyond the known lands stretch vast regions of mathematics still to be discovered.

——————— Maps of a Different Color ———————

A deceptively simple mathematical problem lurks within the brightly colored maps showing the nations of Europe or the patchwork of states in the United States. It's the sort of problem that might trouble frugal mapmakers who insist on painting their maps with as few colors as possible. The question is whether four colors are always enough to fill in every conceivable map that can be drawn on a flat piece of paper so that no countries sharing a common boundary are the same color.

Several conditions turn this mapmakers' conundrum into a well-defined mathematical problem. A single shared point doesn't count as a shared border. Otherwise, a map whose countries are arranged like the wedges of a pie would need as many colors as there are countries. Also, countries must be connected regions; they can't have colonies scattered all over the map (*see* Figure 1.1, *top left and middle*). In addition, the unbounded area surrounding a cluster of countries and filling out the rest of the infinite sheet counts as a separate country that must also be colored.

The four-color problem has intrigued and stumped professional and amateur mathematicians alike ever since it was first suggested in 1852 by Francis Guthrie, who noticed he needed only four colors to color the counties on a map of England. In a letter to his younger brother, Frederick, Guthrie asked whether it was true of *any* map. Frederick in turn described the problem to his mathematics instructor, the British mathematician Augustus DeMorgan (1806–1871). The problem intrigued DeMorgan, and he quickly appreciated that it wasn't quite as simple to solve as it appeared at first glance.

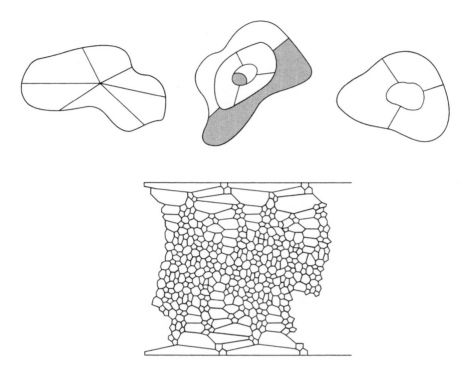

Figure 1.1 *Top left*: A single shared point doesn't count as a shared border. *Top middle*: Countries can't have colonies. *Top right*: Why four colors are needed. *Bottom*: Try coloring this one with just four colors.

Three colors are certainly not enough for every possible map. It's easy to come up with a simple example that requires four colors (*see* Figure 1.1, *top right*). Four colors turn out to be necessary in any situation in which a region has common borders with an odd number of neighboring regions. A map showing the states of the United States features several such instances. For example, Nevada has borders with five states and Kentucky borders with seven. Not counting the water but including islands, Rhode Island has boundaries with three states—Massachusetts, Connecticut, and New York. In each case, a fourth color is always necessary to color the map.

It's not at all obvious, however, whether a fifth color is needed for more complicated maps. Indeed, it's not sufficient to show that one cannot have five countries that each shares a border with the other four. Even if a map contains no such set of five countries, it might still require five colors. Mathematicians could prove, however, that no more than five colors are always sufficient to color any map.

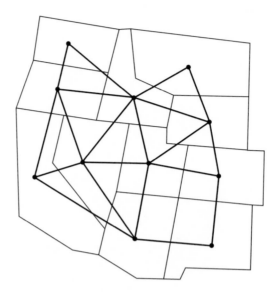

Figure 1.2 Any map on the plane can be converted into a planar graph. In this example involving a stylized map of the western United States, lines join points (nodes) within any two states that share a boundary.

To get a sense of how many colors are needed, one can experiment by drawing and coloring various configurations. It readily becomes apparent that the shape of adjacent regions doesn't matter, but their arrangement does. Indeed, when mathematicians work on the problem, they draw graphs—networks of lines and points—to represent different configurations. That involves putting a point (often called a vertex or node) at each country's capital and drawing a line (described as an edge) to connect the capitals of any pair of countries that has a common border (*see Figure* 1.2).

The four-color problem defied solution for more than a century, despite years of map sketching, graph drawing, and logical argument. In 1976, Kenneth Appel and Wolfgang Haken of the University of Illinois announced they had finally proved the four-color theorem. It was the kind of news that mathematicians greet with champagne toasts, but there was a shock awaiting anyone curious about how the four-color theorem had been conquered. Hundreds of pages long, the proof was truly intimidating. Even more dismaying to many mathematicians was that, for the first time, a computer had been used as a sophisticated accountant to enumerate and verify a large number of facts needed for the proof. There was no simple, direct line of argument leading to a readily verifiable conclusion.

The computer allowed Haken and Appel to analyze a large collection of possible cases, a collection shown by mathematical analysis to be sufficient to prove the theorem. That task took 1,200 hours on one of the fastest computers available at the time; it would have taken practically forever by hand. It also meant the proof could not be verified without the aid of a computer. Furthermore, Appel and Haken had tried various strategies in computer experiments to perfect several ideas that were essential for the proof.

Even now, doubts still linger about the validity of the proof presented by Haken and Appel. Although most mathematicians accept the idea that the Haken-Appel approach works, persistent rumors insist that there is something wrong with the proof. Some people hint that the computer program was faulty, and others that the method itself is wrong in some way.

In 1996, Neil Robertson, Daniel P. Sanders, Paul Seymour, and Robin Thomas presented a new proof of the four-color theorem. Robertson, a mathematician at Ohio State University, and his colleagues had started out trying to verify the Haken-Appel proof but soon gave up. They decided it would be more profitable to work out their own proof, following a somewhat streamlined version of the approach taken by Haken and Appel. Instead of checking the 1,936 graphs (later cut to 1,476 configurations) required by the original proof, they reduced the number to 633 special cases and proceeded from there. Still, the new proof contains computer steps that can't be verified by humans.

"However, from a practical point of view, the chance of a computer error that appears consistently in exactly the same way on all runs of our programs on all the compilers under all the operating systems that our programs run on is infinitesimally small compared to the chance of human error during the same amount of case-checking," the mathematicians insist. "Apart from this hypothetical possibility of a computer consistently giving an incorrect answer, the rest of our proof can be verified in the same way as traditional mathematical proofs."

Appel, who later moved to the University of New Hampshire, once remarked: "It has troubled our profession that a problem that can be understood by a school child has yet to be solved in a way that better illuminates the reason that only four colors are needed for planar maps." Although the theorem is true, it still awaits a simple, incisive proof that does not require hours of computer time.

The proof of the four-color theorem may also represent a new and intriguing facet of mathematics. Although a shorter, more elegant proof may someday be found, it's also possible that no such proof exists. The theorem may be one for which there never will be a proof short enough to be readily understood at a glance. It's an example of a curious, remarkably common occurrence in mathematics—the coupling of a succinct, easily understood problem with an incredibly complicated proof.

Math Lands

Two decades after Haken and Appel's pathbreaking proof of the four-color theorem, computers have become commonplace tools used by nearly all mathematicians, not only to communicate with each other but also to explore mathematical questions. At first, the change came about slowly. In 1986, for instance, Stanford University's Joseph Keller wryly remarked that at his institution, the mathematics department had fewer computers than any other department, including French literature. Many mathematicians had the feeling that using a computer was akin to cheating and insisted that computation was merely an excuse for not thinking harder.

That sort of reasoning persists in some mathematical circles, but most mathematicians now consider the computer an essential part of their work. By providing vivid images that suggest new questions and by allowing a wide variety of calculations of test cases, computers are helping to mend a rift that had developed more than a century ago between pure and applied mathematics. These changes are enriching a discipline that outsiders sometimes regard as an abstract, even useless, pursuit.

To many people, mathematics—unchanging, reliable, dusty with age—has an aura of authority and rests on a firm foundation of pure logic. It promises certainty. Schoolchildren learn strict, seemingly infallible rules, ranging from the mechanics of addition to the intricacies of factoring algebraic expressions. Engineers routinely calculate specifications, consulting reference volumes filled with mathematical formulas. Stockbrokers use the logic built into elaborate computer programs to help make quick decisions about when to buy and sell.

Behind the apparently stolid, pristine, immutable public face of mathematics, however, lies the exciting, turbulent, ever-changing world of mathematical research. Just as physics and other sciences go through episodes of both revolution and evolution, mathematics, too, changes and grows, not only in the way it is applied but also in its fundamental structure. New ideas are introduced; intriguing connections between old ideas are discovered. Chance observations and informed guesses develop into whole new fields of inquiry.

The territory, or continent, of mathematics can be divided into three large chunks. The first, called algebra, involves the study of systems of symbols. In general, an algebra consists of a number of mathematical entities (such as integers, matrices, vectors, or sets) and operations (such as addition or multiplication) with formal rules expressing the relationships between the mathematical entities. It includes, for example, the rules needed for adding, subtracting, multiplying, and dividing the 1s and 0s (or binary digits) that zip through a computer's microprocessor and reside in its memory.

Symbol systems and the operations that are performed within them can be classified in much the same way that animals can be divided according to genus and species. Just as zoologists can place cats in the family of mammals, mathematicians have places for algebraic systems that obey certain rules. One particularly important and useful category is called the group, which shows up as an organizing principle in all branches of mathematics and in crystallography, particle physics, and other sciences.

Analysis, the second major piece of the mathematical continent, concerns functions, which express relationships. Simply put, a function is any rule that assigns a fixed output to a given input. For example, if the function is squaring, the number 3 is paired with the number 9 (the result of multiplying 3 by itself), and the number 7 is paired with the number 49. The development of calculus, worked out independently by Isaac Newton (1642–1727) and Gottfried Leibniz (1646–1716), is one of the central achievements of analysis.

The third region, geometry, is the study of the properties of shapes and spaces. Most people are familiar with the rigid forms of Euclidean geometry—squares and cubes, circles and spheres, congruent triangles and parallel lines. But geometries can take on many different guises, reaching into higher dimensions and obeying varied rules. Topology focuses on geometric features that remain unchanged after twisting, stretching, or otherwise deforming a geometric space. Problems such as coloring maps, distinguishing knots, and classifying surfaces, or manifolds, not just in one, two, and three dimensions, but in higher ones as well, all fall within topology.

Lying a short distance offshore from the mathematical mainland are the islands of number theory and set theory. Number theory, at one time considered the purest of pure mathematics, is simply the study of whole numbers, including prime numbers—those numbers evenly divisible only by themselves and 1. Once the playground of a few mathematicians fascinated by the curious patterns and properties of numbers, this field now has considerable practical value. Identifying primes and finding the prime factors of a composite number play a crucial role in many cryptographic schemes— systems designed to keep secrets secret.

A set is any collection of entities, including numbers, that belong to a well-defined category. A "dog," for example, is a member, or element, of the set of all "four-legged animals." Numbers such as 57, −4, and 6,897 belong to the set of integers, whereas numbers such as 7/11, 4.67, and π (pi), the ratio of a circle's circumference to its diameter, belong in other categories. Set theory concerns the study of the structure and size of sets, as defined by various axioms and rules.

Somewhere nearby lie the land masses of statistics and computer science. Both have close ties with mathematics, and the links are becoming increasingly important. Computer science can be thought of as the study of algorithms—the methods or step-by-step procedures used to solve given classes of problems. In cooking, a recipe is the algorithm that guides the cook in transforming a motley collection of ingredients into a scrumptious cake. Mathematicians also need recipes. In classical geometry, the ancient Greeks devised a slew of procedures, employing only ruler and compass as tools, for performing a variety of geometric feats, including the bisecting of angles and the drawing of regular figures such as hexagons. Later mathematicians spent much of their time looking for algorithms for efficiently computing π, finding logarithms, identifying primes, and performing countless other mathematical tasks.

Today's explosive growth in computer use adds urgency to the investigation of algorithms. Because computers operate on the basis of a small, built-in set of operations, the programs that instruct them, which are essentially algorithms, must be made as efficient and reliable as possible. The mathematical analysis of algorithms has spawned the field called computational complexity. Researchers try to measure how the number of steps needed to solve a problem varies as the size of the problem increases and as the details of the problem change, whether it involves finding the shortest route linking a certain number of cities or determining how many different ways items can be snugly packed into a set of bins of given capacity.

Mathematical research in modern times is an extensive enterprise. Because thousands of mathematicians publish hundreds of thousands of pages of new mathematical findings every year, most mathematicians find it difficult to keep up with what's happening across the field. Most of the time, they have to be content with being up-to-date in only a narrow furrow in the field. Consequently, mathematics tends to stay tightly packed into segregated compartments.

Nevertheless, many of the most striking mathematical results of the last few decades involve notions developed in one field that turn out to be a key element in solving outstanding problems in another, seemingly unrelated field. Such mathematical transfers have occurred in number theory, for example, where ideas about so-called elliptic curves proved crucial in the invention of an improved method of factoring integers into their prime components (*see Chapter* 2) and in the proof of Fermat's last theorem (*see Chapter* 8). The many such links that constantly come to light attest to the essential unity of mathematics.

———— — – - — Logging On —— —— — ——

The availability of very fast computers with large memories has led to two important developments in science and technology, both of which rely heavily on mathematics. The first development was the use of mathematical equations to represent a physical situation, such as air rushing over an airplane's wing or the chemical reactions leading to the formation of ozone. When they are good enough, such mathematical models allow researchers to replace many of their wind-tunnel and test-tube experiments with computer manipulations.

The second development arose because computers are invaluable for extracting relationships and patterns hidden deep within enormous amounts of data. For example, they can sift through seismic data to pinpoint the location of an oil field or analyze X-ray absorption measurements to determine the site of a brain tumor. The processing of data and the simulation of physical systems require subtle, sophisticated techniques. More often than not, a piece of mathematics worked out years before—and believed to be totally without practical value—finds a role in the "real" world.

One particularly striking example of the value of computer simulations and of the interaction between mathematics and science is in the work of Charles Peskin, an applied mathematician at New York University's Courant Institute of Mathematical Sciences. He has spent more than two decades perfecting a computer model of blood flow in the heart. Peskin's aim is to create a tool that physiologists can use to illuminate how real hearts function (or malfunction) and that designers of mechanical heart valves can use to test prototypes without having to rely entirely on animal experiments and clinical trials with human patients. In effect, he brings equations to life.

The heart is a complicated mass of tissue, consisting of bundles of oriented muscle fibers. Some fibers in the heart wall wind around to form shells that look like nested doughnuts. Other fibers take more complicated paths that lace together the heart's right and left halves. Peskin initially concentrated on modeling the heart's left side and on the movement of one particular valve, known as the mitral valve. With every heartbeat, the mitral valve's thin, flexible flaps of tissue smoothly slip out of the way when blood pushes forward. When the heart contracts, the valve snaps shut to keep blood from flowing back the wrong way.

Peskin's original two-dimensional model included the characteristics of both the blood flow and the heart chamber's muscle tissue. The blood was represented by a large number of discrete points, each with a specific velocity and pressure. These velocities and pressures change as the fluid elements interact with their neighbors in ways that can be calculated using equations from physics. The heart's muscle fibers were modeled by a collection of moving particles joined by tiny springs. The properties of these elastic links change over time to simulate the change in tissue stiffness during a heartbeat. Like an elastic band, the modeled tissue responds flexibly to the blood's pressure while it also exerts a force on the flowing blood.

The net result of the millions of computations required for Peskin's model was a sequence of pictures that could be strung together to produce a dramatic movie of a beating heart (see Figure 1.3). The vivid two-dimensional images clearly showed where the flow was uneven, how differently shaped artificial valves responded, and even the likely performance of a weakened or diseased natural valve. Peskin and collaborator David

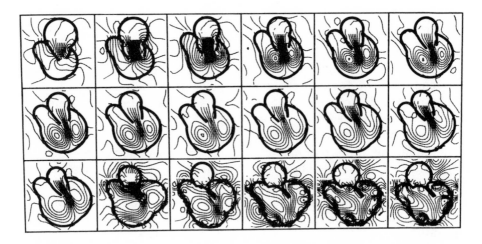

Figure 1.3 In this time sequence from a two-dimensional simulation of the human heart, computed streamlines show how blood flows through the heart's mitral valve.

McQueen, a mechanical engineer, used the model to design a novel artificial heart valve, which has been patented and licensed.

The human heart, however, is a three-dimensional organ, and moving the simulation to three dimensions raised interesting and difficult geometric questions, particularly when it came to depicting the blood's computed flow pattern within a three-dimensional structure. The improved model heart consists of hundreds of closed curves representing muscle fibers—as if the heart were constructed out of a huge array of rubber bands. Mathematically, as in the two-dimensional case, each curve is built out of points, with each pair of adjacent points joined by a spring. A computer keeps track of these points and those representing blood. As the calculation proceeds, the numerical heart beats and simulated blood sloshes through the organ (*see Figure* 1.4).

Coupling flexible boundaries with a moving fluid made up of particles was perhaps one of the greatest mathematical challenges that came up during the development of the heart model. The mathematical and computational methods that Peskin and his collaborators eventually developed to solve the problem can now be applied in many other situations—from the movement of fish in water and the passage of waves through the inner ear to the study of flow in thin-walled blood vessels and the design of aerodynamically efficient sails and parachutes.

To an increasing number of practitioners, computer simulations rooted in mathematics represent a third way of doing science, alongside theory and experiment. In the past, a physical theory often consisted of a set of

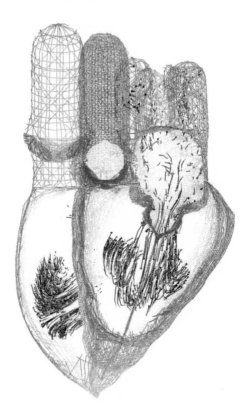

Figure 1.4 Cutaway view of a three-dimensional computer model of the heart. Blood flow is indicated by streamlines, which trace the trajectories of particles immersed in the flow. The current position of each such particle is shown as a small blob, and the recent past positions are shown by a fading tail.

equations. A theory today may consist of a computer program that models the way a system is supposed to evolve. Once experience shows that a given computer program accurately simulates a real system, then experiments may be done using a computer instead of in the lab. Of course, the computer model may produce results that fail to mimic reality. Then it's back to the keyboard to fix the model.

The use of computer models has spread rapidly to fields as diverse as astronomy, chemistry, and psychology and is illuminating the dynamics of pinwheel galaxies, the tremors that shake protein molecules, and the mechanisms that underlie human memory. In many cases, mathematical models of physical situations save time and money. In aerodynamics, for example, typing in a few values at a computer terminal and seeing the results displayed on a color screen can save days otherwise spent crafting a wooden or metal model and testing it in a wind tunnel. Laboratory experiments and observations are still necessary, but now the researcher can choose which experiments are most likely to be useful and may even have a better idea of what behavior to observe and which variables to measure in a given experiment.

When an experiment produces masses of data, mathematical and statistical techniques are available to help researchers make sense of the jumble. The development of X-ray tomography to produce CAT scans and of other methods for probing the human body provides a good example of how much mathematics, both old and new, is needed to make such techniques work.

In computerized X-ray tomography, hundreds of X-ray beams, finer than needle points, zip through a slice of the human body. As each beam passes through tissue, bone, and blood, it weakens by an amount that depends on what it encounters. In mathematical terms, firing an X-ray beam through an object and seeing what happens to the beam is equivalent to finding the projection of a function along a given line. This mathematical operation is called a Radon transform, named for the Bohemian mathematician Johann Radon (1887–1956), who worked on such mathematical manipulations early in the twentieth century. Radon's work appears to have been pure mathematics carried out for its own sake.

In the case of tomography, researchers must sift through their X-ray absorption data to compute the tissue density at a particular spot. In other words, from patterns hidden within all the X-ray intensity measurements, they need to work out the tissue arrangement that gives the observed intensities. That's not unlike looking at a complex pattern of ripples on a water surface and trying to decide where the ripples started and how many sources there were. Mathematically, the operation means computing the original values when only a set of averages is known. It means performing a Radon transform in reverse.

Radon proved the basic theorems that specify under what conditions a function can be reconstructed from various projections or averages. Radon's ideas, however, apply only to continuous functions, which can be represented as smooth curves. In tomography, an infinite number of X-ray beams would have to be used to sample the entire cross section before the tissue densities at every spot could be accurately computed. Because only a finite number of X-ray measurements are actually made, mathematicians have had to work out approximate methods that allow cross sections of human organs to be reconstructed with a minimum of error. They try to make sure that an error isn't exactly where a tumor may be. New mathematical work is leading to faster algorithms so that X-ray images can be generated almost instantaneously.

Work on the discrete Radon transform and its inverse as applied to tomography has also suggested some interesting mathematical questions. The mathematicians Ronald Graham and Persi Diaconis have taken a close look at what can be deduced in situations in which certain averages are known but the original data are missing. Graham wondered, for example, to what extent secret data contained in confidential files can be uncovered if the right questions are asked. Cracking a confidential database can be likened to the old parlor game of twenty questions. A player receiving only yes-or-no answers yet asking the right sequence of pertinent questions can often deduce the identity of some hidden object or person. In the same way, the answers to a series of general questions addressed to a particular database can add up to a revealing portrait of something that is supposed to be secret.

A simple example illustrates how such a scheme might work. Suppose someone wants to find out Alice's salary. The inquisitor has access to information revealing that the average of Alice and Bob's salaries is $30,000; the average of Alice and Charlie's salaries is $32,000; and the average of Bob and Charlie's salaries is $22,000. This provides enough information to deduce that Alice's salary is $40,000.

Researchers often face a situation in which certain averages are known but the original data are missing. If eight data points happen to be identified with the eight vertices of a cube and each of the eight numbers is the average of its three nearest neighbors, it's possible to deduce the actual but currently hidden value associated with each vertex (*see Figure* 1.5). In this situation, the actual value at each vertex is equal to the sum of the nearest-neighbor averages minus double the average at the corner farthest from the point of interest. Curiously, the point that makes the largest contribution to the answer is the one that's farthest away.

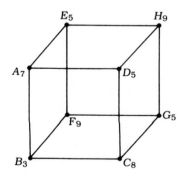

Figure 1.5 The actual value at vertex A is the sum of the averages shown at B, D, and E minus twice the average at G: $(3 + 5 + 5) - 2(5) = 3$. The values of the other vertices are 6, 3, 6, 9, 3, 12, and 9.

Diaconis and Graham have developed a mathematical theory, based on the idea of discrete Radon transforms, that helps to decide how many and which averages are needed to crack a database or to analyze statistical data. At the root of their exercise is the mathematical concept of how completely a bunch of averages captures the mathematical relationship underlying a data set.

Pure Power

Although mathematical methods play important roles in statistics and computer science, the focus in these fields is on accomplishing particular goals in an efficient, practical manner. In contrast, pure mathematics delves more deeply into the existence and nature of mathematical objects, even when the objects can't be explicitly computed or constructed. Mathematics also involves proof. Once proposed, mathematical conjectures, or guesses, go through a trial by fire before they emerge as carefully defined theorems with a permanent place in the structure of mathematics.

Nonetheless, even what is traditionally called pure mathematics isn't immune to experiment and observation. To explore the properties of prime numbers, mathematicians centuries ago compiled lengthy tables, using them to look for trends, to guess which properties such numbers have, and to see how primes are distributed among all whole numbers. When the great mathematician Carl Friedrich Gauss (1777–1855) was 15 years old, for example, he obtained a list of all prime numbers less than 102,000. The young Gauss spent hours counting the number of primes in blocks of 100, 1,000, and 10,000 consecutive integers. He found that as numbers get larger, the proportion of primes decreases in a particular way, and later, he conjectured that the number of primes smaller than a given integer is approximated by a mathematical expression involving logarithms (*see Chapter* 2).

Today, computers facilitate this kind of list making and sorting to identify trends and patterns. Such computations, although not always an integral part of the final proof, often suggest what ought to be proved—or even conjectured in the first place.

Numerical experiments, in which computers are used as untiring accountants and bookkeepers, have already suggested important ideas about the behavior of algebraic expressions and differential equations. One result of such experiments is the discovery of chaotic regions coexisting with islands of stability among the solutions of equations describing planetary motion and other physical phenomena. Out of this comes the disturbing news that some mathematical procedures—for example, those used to solve equations—may not be as reliable or stable as people had thought (*see Chapter* 6).

Computer visualization and physical observation have proved highly effective in advancing geometry. Experiments with soap films show the tremendous variability in the shapes of surfaces. Computer-generated pictures of four-dimensional forms reveal unusual geometric features. The crinkly edges of coastlines, the roughness of natural terrain, and the branching patterns of trees point to structures too convoluted to be described as one-, two-, or three-dimensional. Instead, mathematicians express the dimensions of these irregular objects as decimal fractions rather than whole numbers.

Experiments and observations, however, are not the whole story. Mathematical ideas, once conceived, seem to take on a life of their own and often wander far from their origins. For example, the abstract concept of minimal surface, inspired by visions of soap films stretched across wire frames, now includes forms that would, as soap films, be too fragile or tortuous to ever be observed in nature.

At the same time, playful mathematics, originally created for its own sake in the course of solving an engrossing puzzle, may more often than not find application in surprising venues. One such case is the startling link that has developed between tiles of certain shapes that slip into place to produce a distinctive nonrepeating pattern and an highly unusual type of crystal that scientists had not expected to find.

——— — — -. The Fivefold Way — ——— —

In its simplest form, a tiling problem in mathematics resembles the practical task of covering a floor with ceramic tiles. Tiles in the shape of equilateral triangles, squares, or hexagons fit together nicely to generate highly symmetric, perfectly regular patterns (*see Figure* 1.6). On the other hand, tiles in the shape of regular pentagons, with five sides of equal length and equal interior angles, can't cover a bathroom floor completely. Attempts to lay such tiles out on a flat surface invariably leave gaps in the pattern. Nonetheless, it's possible to lay down tiles of two different shapes to create a pattern having the fivefold symmetry of a regular pentagon.

In 1974 the mathematical physicist Roger Penrose of Oxford University was engaged in a search for the smallest possible set of different tiles that, used together, would cover a surface without forming a regularly repeating pattern. He discovered that he could do it with two tile shapes. Such a pair can be created by cutting a rhombus, a figure resembling a skewed square or a diamond, into two pieces in a special way (*see Figure* 1.7). One piece of the rhombus resembles an arrowhead and is called a dart. The other piece, whose blunter end fits snugly into the arrowhead's notch, looks like a diamond with one foreshortened end. It is known as a kite.

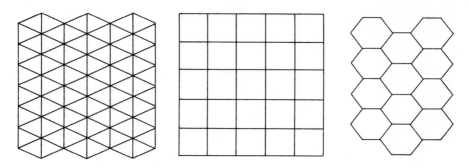

Figure 1.6 Regular polygons, such as equilateral triangles, squares, and hexagons, can be arranged into symmetric tilings.

Not surprisingly, when tiling the plane, it's possible to reassemble kites and darts into rhombuses and to arrange these forms in such a way as to create a regularly repeating, or periodic, arrangement. However, by working out rules that specify which edges of one tile can fit with certain edges of another, one can force the tiles to fit together only in certain ways, generating a nonrepeating pattern. That can be done by decorating the appropriate sides or angles of each dart and kite with colored bands or by using small, interlocking tabs like those on jigsaw-puzzle pieces (*see Figure* 1.8). Such matching rules

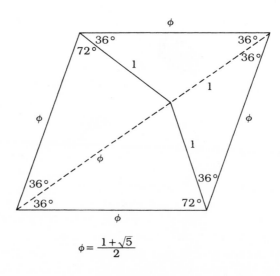

Figure 1.7 Dividing a rhombus in a manner based on the golden ratio, ϕ, produces a kite and a dart.

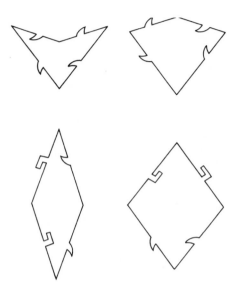

Figure 1.8 Adding notches and projections ensures that tiles cannot be put down in a periodic pattern.

forbid juxtapositions that lead to periodic arrangements. Remarkably, the resulting patterns have an approximate fivefold symmetry (*see* Figure 1.9).

As it turns out, there are many different pairs of quadrilateral shapes that form a nonperiodic tiling pattern, although all are related in some way to the original kite-and-dart pair. One particularly useful set consists of a pair of diamond-shaped figures—one fat and one skinny (*see* Figure 1.10). The two types of tiles combine to form intriguing arrays of unusual symmetry (*see* Figure 1.11 *and* Color Plate 1).

Penrose didn't have anything practical in mind for the remarkable tiling patterns he created when he started drawing his diamond tapestries. The exercise was simply a challenging mathematical game that tickled his fancy. It took the discovery in 1984 of tiny metallic crystals composed of aluminum and manganese to alert the scientific world to Penrose's remarkable tilings. The crystals had a form as startling and unexpected as a five-pointed snowflake. X rays and electrons reflected from the crystals created an image consisting of concentric rings of well-defined spots arranged so that the overall pattern had a fivefold symmetry—an occurrence, according to conventional theory, that shouldn't have been possible (*see* Figure 1.12).

Formulated more than a century ago, the standard rules of crystallography rigidly maintain that the building blocks of a particular crystal consist of identical arrangements of atoms, ions, or molecules. Constructing a crystal out of these units is much like laying bricks to build a wall.

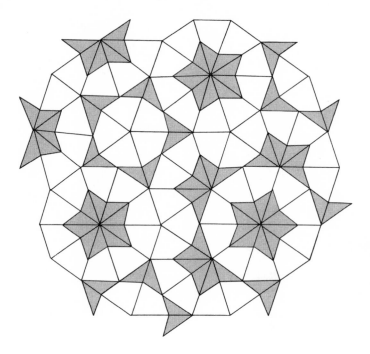

Figure 1.9 An example of kite-and-dart tiling.

This requirement for order in a crystal suggests that only triangular, square, or hexagonal lattices and variations on these forms have the necessary regularity. In common salt, for example, sodium and chloride ions sit at the corners of cubes, and these cubes stack neatly to fill out each salt crystal. As a result, one orderly row of ions succeeds another at regular intervals.

According to the well-established suppositions of crystallography, formalized in the branch of mathematics known as group theory, only a small list of rotational symmetries is possible for crystals. For instance, a crystal

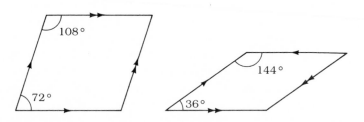

Figure 1.10 A pair of rhombuses—fat and skinny diamonds—can also be used to tile a plane nonperiodically.

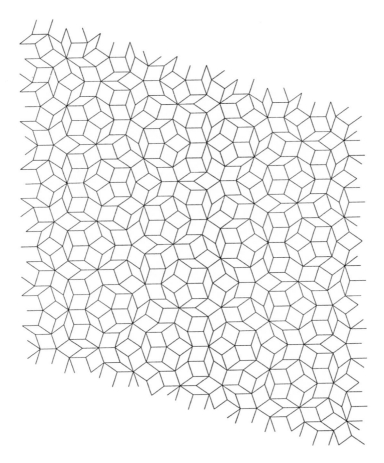

Figure 1.11 Example of a Penrose tiling made up of thin and fat rhombuses.

is said to have threefold rotational symmetry if the crystal latticework would appear unchanged after the crystal is rotated 120 degrees about one of its axes. In general, crystals can have twofold, threefold, fourfold, or sixfold axes of rotational symmetry. Nonperiodic structures are forbidden.

In the case of the novel manganese-aluminum alloy, the sharpness of spots in the X-ray diffraction pattern indicated that the atoms are organized in some orderly fashion rather than randomly placed. Nonetheless, the pattern's unusual fivefold symmetry suggested that the atoms can't lie in one of the orderly arrangements conventionally assigned to crystals. Instead of repeating throughout the structure at some regular interval, atoms of a so-called quasicrystal appeared to be clustered in a highly unusual geometric arrangement.

It was natural for physicists and others to turn to Penrose tilings and their three-dimensional analogs as reasonable models of the basic units

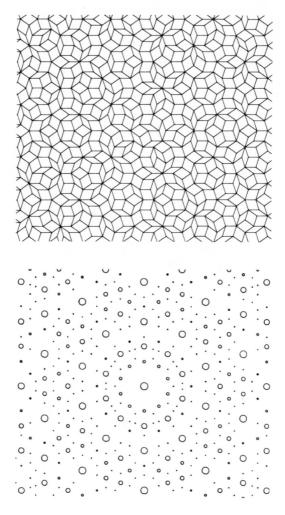

Figure 1.12 A Penrose tiling (*top*) produces a diffraction pattern of spots (*bottom*) that has a fivefold symmetry.

that might fit together to produce a quasicrystal. Indeed, the Penrose tiling embodied several features characteristic of quasicrystal structures. It suggested, for example, that the atoms of a quasicrystal organize themselves into two types of clusters, which act as building blocks, rather than into a single type of block typical of ordinary crystals.

The trouble is that a Penrose tiling requires a set of rules that specifies how the tiles must be placed edge to edge. It's hard to imagine how clusters of atoms in quasicrystals could interact in ways that mimic such complicated matching rules. Moreover, it's difficult for anyone to piece together, say, a hundred Penrose tiles without error, but perfect quasicrystals made up of 10^{23} atoms can form in minutes.

The inadequacies of the original Penrose tiling model prompted searches for alternative ways of describing quasicrystals. Petra Gummelt of the University of Greifswald in Germany was one of the first to come up with such a scheme. She used a ten-sided, or decagonal, tile as her basic structural unit. Rather than abutting each other like Penrose tiles, the decagons overlap in specific ways. The lumpiness one would get with real bathroom tiles was not a problem, because her mathematical tiles were only two-dimensional.

When Paul J. Steinhardt of the University of Pennsylvania first heard of Gummelt's pattern, he was skeptical. It wasn't clear to him from her report that the construction actually worked. However, he and colleague Hyeong-Chai Jeong of the University of Maryland ended up confirming that Gummelt was correct, and they worked out a simpler version of her original proof that illuminated the link between her decagons and Penrose's diamond tiles (*see Figure 1.13 and Color Plate 2*). They then proved that her overlap rules are equivalent to Penrose's matching rules.

The improved mathematical model eventually developed by Steinhardt and Jeong may shed light on the interactions responsible for quasicrystal formation. Their model requires only a single type of building block for constructing a two-dimensional quasicrystal. A sensible way of interpreting the overlap rules in physical terms is as a sharing of atoms by neighboring clusters, rather than as two clusters penetrating into each other. Such a possibility is consistent with experimental data on the positions of clustered atoms in quasicrystals.

Extended to three dimensions, this approach could provide a simple, unified picture of how both ordinary crystals and quasicrystals form. Indeed, the results show that the atomic structure of quasicrystals and ordinary crystals can be understood in terms of a single repeating unit.

The discovery of quasicrystals also prompted a reevaluation of what constitutes a crystal. Scientists in the past generally defined crystals as materials whose atoms are arranged in periodic patterns, as if the atoms were locked into specific positions on a regular grid. Many researchers are now beginning to think of a crystal as simply any solid that yields a diffraction pattern consisting largely of well-defined, bright spots, as recorded on a photographic plate when X rays or electrons pass through the material. The new definition greatly enlarges the number of geometric arrangements that can be considered crystalline. It also raises a host of subtle and difficult mathematical questions concerning the relationship between specific geometries in abstract mathematical models of crystals and the positions and intensities of spots in diffraction patterns.

In quasicrystal geometry, mathematicians and physicists share the experience of seeing in its complexity a pattern that is subtle yet powerful and a

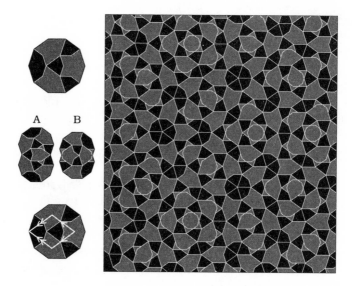

Figure 1.13 A nonperiodic tiling can be generated by overlapping a single type of decagon tile (*left top*). A tile can overlap its neighbors if the overlapping regions have the same shading (*left middle*). Such a covering of the plane with decagons (*right*) can be transformed into a Penrose tiling by inscribing a fat Penrose diamond (*left bottom*) inside each decagon. The arrows on the diamond are an alternative to lines as a way of specifying the Penrose matching rules. Adjacent tiles must have arrows of the same direction and number on adjoining edges.

symmetry that is elusive yet enthralling. Physicists see in these patterns hints of how nature may organize matter under certain circumstances. They invent and invoke special rules to create models that appear to mimic the behavior of real-life quasicrystals. Mathematicians enjoy the opportunity to explore the diversity of forms arising from simple rules and principles. They look for links between these patterns and other types of mathematics. They try to enumerate the geometric possibilities in any given situation. They sense the mystery.

Even a brief glimpse of modern mathematical research reveals a dynamic enterprise of provocative ideas and questions. Far from being a domain of largely settled issues, mathematics is truly a wilderness. The well-mapped settlements lie few and far between, scattered across the continent and linked by a still skimpy network of highways and trails, some better traveled than others.

It's an exciting adventure to explore some of the newest paths that penetrate the mathematical wilderness—to pursue primes, to untangle twisted spaces, to delve into higher dimensions, to wander in labyrinths, to battle with chaos, and to puzzle over tiling patterns. These images and notions may seem fantastic and metaphorical but they're not. They are the objects studied by mathematicians—objects that we will see more of on our tour. Novel landscapes, new vistas, and unexpected pathways lure the mathematical tourist to the frontier, where questions and conjectures abound.

2

Prime Pursuits

A young child learning to count can get to 20 by enumerating fingers and toes. Higher numbers take on meaning as the child's experience grows. Mathematicians also find meaning in numbers. They start with patterns or relationships displayed by relatively small counting numbers and try to find out whether the patterns hold for ever larger ones. On such a foundation, they have built the discipline known as number theory.

In recent years, number theory has come out of its mathematical closet to play a crucial role in public affairs. The playful work of long-ago mathematicians turns out to have key applications in cryptography and computer security. When you slip your ATM card into a cash machine, you take advantage of the convenience and safety offered by cryptographic schemes based on manipulating long strings of digits.

—————— Heads or Tails? —————

Alice lives in New York City, and Bob lives in Los Angeles. They are recently divorced and communicate, when they do at all, only by telephone. Now they have to decide who should get an antique table that they have just jointly inherited. They agree to toss a coin. Neither one of them, however, would like to choose, say, heads and then hear the other person exclaim over the telephone, "Here goes . . . I'm flipping the coin . . . you *lose!*" Somehow, Alice and Bob must find a way of tossing the coin without suspecting each other of cheating.

The dilemma confronting Alice and Bob is related to a problem faced by many computer users. In the high-tech world of telephone-linked and networked computers, electronic mail, the Internet, and the World Wide Web, methods are needed to guarantee that secret messages arrive intact, that funds are transferred securely and correctly, and that contracts and agreements are properly signed. Otherwise, electronic letters are as open as postcards, and a person may be tempted to lie or cheat to gain an advantage over a competitor.

One ingenious scheme for remote coin tosses blends the capabilities of modern high-speed computers with some basic mathematics rooted in ancient Greece. In essence, this particular procedure converts a call of heads or tails into the problem of factoring a large number. To win, the caller must find the two smaller numbers that the coin flipper multiplied together to generate the large number. This coin-toss algorithm not only provides Alice and Bob with a way to settle their dispute but also suggests protocols for ensuring the fairness of computer-aided business transactions, such as signing contracts and sending certified mail.

At the heart of the coin-toss algorithm is a procedure called the oblivious transfer, devised by Harvard University's Michael Rabin. The transfer is somewhat like a simple game played with a locked box requiring two different keys. A sender transfers the locked box to a recipient, who finds one key that does half the job. The sender has both keys and, without seeing what the recipient has done, must now pass on one of the two keys. Depending on which key is sent, the recipient will either succeed or fail to open the box. Although the sender's choice controls the outcome, the sender never knows which choice to make to guarantee a particular result.

The oblivious transfer depends on two crucial mathematical ideas. First, mathematicians and computer scientists know several quick ways to determine whether a large number of, say, 100 digits, is a prime number, that is, evenly divisible only by itself and by the number 1. A properly programmed personal computer can do this in a fraction of a second. Mathematicians also believe that factoring a 200-digit or larger composite number, especially when the number is the product of just two approximately equal primes, can take years, even on the fastest available computers. The claim that factoring is intrinsically hard has not yet been proved, but mathematicians have assembled convincing evidence to support their belief. The security of Rabin's transfer hinges on the difficulty of factoring large numbers.

As in several modern cryptographic systems for sending secret messages, modular arithmetic plays an important role in the oblivious transfer. It's the kind of arithmetic that people use every day when they think about clocks and time. Adding 12 hours to the time shown on a clock face brings the clock's hands (or digits) right back to where they were at the start. Moreover, as far as the clock is concerned, adding 14, 26, or 38 hours is the same as adding just 2 hours.

In modular arithmetic, only remainders left over after division of one integer by another are saved. In the clock example, dividing 12 into 14, 26, or 38 produces a remainder of 2. Any other numbers that happen to come up during the computation are tossed out. More formally, given two integers, a and n, the remainder of a divided by n is written as $a(\bmod n)$ and described as "a modulo n." In this arithmetic system, $7(\bmod 5)$ is 2, $13(\bmod 5)$ is 3, and so on. A negative number, $-a$, is treated in the same way as $(n - a)(\bmod n)$. Thus, $-2(\bmod 5)$ is the same as $(5 - 2)(\bmod 5)$ or $3(\bmod 5)$, which equals 3.

Modular arithmetic is a handy key for locking up secrets. Its advantage is that it affords a kind of one-way trapdoor: What goes in doesn't easily get out again. Computing something like 14 modulo 12 is straightforward. The answer is 2. Reversing the operation is trickier. If the remainder happens to be 2 modulo 12 (or 2), the original number could have been 2, 14, 26, or any one of innumerable other possibilities.

Here's how the oblivious transfer would work in the case of a remote coin toss. Alice and Bob use their linked personal computers to perform the toss. Normally, the computers would deal with 100- and 200-digit or larger numbers and probably go through the procedure in less time than it takes someone to read this example. However, to make sure that most of this chapter isn't taken up with lengthy strings of digits, let's say Alice selects much smaller numbers than those a computer would handle.

Alice, in New York, starts the coin toss by selecting two prime numbers, 7 and 13. She multiplies the numbers together and sends the product, 91, to Bob. She keeps the numbers 7 and 13 secret. Bob wins the toss if he can factor 91, that is, find 7 and 13.

Suppose that Bob, in Los Angeles, can't factor 91. First, he can check whether Alice is cheating. She could have sent a number that can't be factored, but Bob's computer has efficient tests to show whether the number is a prime. Once his computer finishes its primality check, Bob randomly selects an integer, say 11, that falls between 1 and 91. The integer may, by chance, turn out to be a factor of 91, and therefore he wins the toss immediately. However, the chances of this happening with a 200-digit number are incredibly small. So Bob squares 11 to get 121. In the modular arithmetic step, he divides 121 by 91 and finds that the remainder is 30. Bob sends this remainder to Alice. He keeps the number 11 secret.

Alice knows the original number, 91, and she looks for all numbers less than 91 that generate a remainder of 30 when squared and then divided by her number. In this case, she can do it by trial and error, but mathematicians use a faster method based on the Chinese remainder theorem, which provides a handy way to reconstruct integers from certain remainders. She finds two pairs: ±11 and ±24. She can send either 11 or 24 to Bob. One of the numbers is Bob's number, but Alice doesn't know which one of the two is his.

If Alice sends the number 11, Bob would have no new information and would lose the toss because he wouldn't be able to factor 91. However, Alice happens to send him 24. Bob adds his own number, 11, to 24 and gets 35. The greatest common divisor of 35 and 91 is 7. This number, 7, will automatically be a factor of 91. Interestingly, the greatest common divisor of 91 and the *difference* of 24 and 11 gives the other factor, 13. Now, Bob can factor Alice's number, and he wins the coin toss. Finding the greatest common divisor is an application of Euclid's algorithm, named for the great Greek geometer and one of the earliest known tools of number theory. The algorithm allows one to find, in a systematic way, the greatest common divisor of two integers without having to factor the two numbers.

The oblivious transfer also makes an appearance in a protocol for sending certified mail by computer. The protocol was developed by Manuel Blum of the University of California at Berkeley. The usual problem with certified mail sent from one computer to another is that although senders get a record that the message was received, they do not get confirmation of the message's content. This allows the possibility of tampering and the appearance of deliberate or accidental errors in the received message.

In this case, the sender, Alice, embeds her message in the digits of 10 large numbers she sends to Bob for factoring. The numbers, each of which is a product of two large primes, are constructed so that Bob must factor all 10 in order to read the full message.

As in the coin-toss example, Bob picks a number at random and runs through the oblivious transfer. Depending on Alice's response, he may or may not be able to factor the first number. Then he goes on to the second number, runs through the algorithm, and so on, until he has tried all 10. After completing this first stage, Bob is likely to be able to factor some but not all of the numbers. In fact, because he has a 50 percent chance of factoring each number, he will, on average, end up with five of the 10 numbers needed. Bob then repeats the procedure, running through all of Alice's numbers in the same order as before but with new random guesses. He continues through these stages until he has enough information to find the factors of all 10 numbers. This usually takes about three or four passes through Alice's list.

Eventually, Bob has the information to find and decipher Alice's message, and Alice has a record of Bob's trial guesses. The collection of Bob's requests constitutes her receipt. If Bob were to deny receiving the message, a judge could determine from the transaction record that Alice provided enough information for Bob to factor the numbers and that Bob must have received the message she sent.

A variation of Blum's certified-mail scheme, with some additional safeguards, can ensure that contracts are signed simultaneously in different parts of the world. Traditionally, businessmen, diplomats, or generals have gathered in one place to sign contracts, agreements, or treaties. Because they don't trust each other, they want to see that the correct signatures are affixed at the appropriate time. Without a protocol and if the parties to an agreement are in different places, someone may be tempted to cheat. All kinds of rituals and ceremonies have evolved over the centuries to make sure that no person or country gains an advantage because of delays in signing a document.

A form of electronic signature is already familiar to people who use a bank's automated teller machines. To withdraw or deposit money, the customer must have a coded card and be able to enter a secret personal identification number. That combination constitutes the signing of a document permitting the transaction.

Blum's protocol is a more sophisticated and secure version of a bank's automated transaction system. Each person involved in the deal sends a copy of the contract to the other person, in the same way that each person would send certified mail by computer. Each person also receives a receipt, which can be interpreted as a signature. The agreement includes a clause stating that the contract is valid only if the participants have in their hands (or more likely stored in their computers) both contract copies and receipts. The contract is valid only if the transaction is completed by both sides.

The protocols described are only two of many schemes proposed to reduce the risk of fraud when people use computers. Some schemes begin as simple games like tossing a coin or playing mental poker and evolve into elaborate protocols for electronic signatures or exchanging secret messages. These protocols often employ fundamental mathematical ideas that were discovered centuries ago without computers or applications in mind.

Prime Properties

The study of prime numbers has long been a central part of number theory, a field traditionally pursued for its own sake and for the mathematical beauty of its results. The number theorist Don Zagier once commented that "upon looking at prime numbers, one has the feeling of being in the presence of one of the inexplicable secrets of creation."

Like the chemical elements in chemistry or the fundamental particles in physics, prime numbers are building blocks in the mathematics of whole numbers. Whole numbers other than primes, known as composite numbers, can be written as the product of smaller primes. In fact, according to the fundamental theorem of arithmetic, each composite number has a unique set of prime factors. Hence, the composite number 20 can be broken down into the prime factors 2, 2, and 5. No other composite number has the same set of factors. The number 1 is considered to be neither prime nor composite. The number 2 is the only even prime number.

Whereas chemists have merely a hundred or so elements to play with, mathematicians interested in number theory must deal with an unlimited supply of primes. More than 2,000 years ago, Euclid of Alexandria (fl. ca. 300 B.C.) proved that there is an infinite number of primes, and mathematicians ever since have been caught up in a never-ending pursuit of these special numbers.

Euclid's argument is instructive. Suppose there is a finite number of primes. That means there's also a largest prime. Multiply all the primes together, then add 1. The new number is certainly bigger than the largest prime. If our initial assumption is correct, the new number can't be a prime. Otherwise, it would be the largest prime. Hence, it must be a composite number and divisible by a smaller number. However, because of the way the number was constructed, all known primes, when divided into the new number, leave a remainder of 1. Therefore, the initial assumption can't be correct, and there can be no largest prime.

Proofs of the existence of infinitely many primes, however, give no indication of how these numbers are distributed or where to find a particular member of the sequence of prime numbers. "Prime numbers have always fascinated mathematicians," Underwood Dudley of DePauw University wrote in a 1978 textbook. "They appear among the integers seemingly at random, and yet not quite: There seems to be some order or pattern, just a little below the surface, just a little out of reach."

The search for patterns and trends plays an important role in the study of the distribution of prime numbers. Clearly, although Euclid proved that the list of primes goes on forever, the stretches between primes, on average, get longer and longer (*see Figure* 2.1). In other words, primes gradually become more scarce as numbers get larger.

There are four primes among the first 10 integers, 25 among the first 100, 168 among the first 1,000, 1,229 among the first 10,000, and 9,592 among the first 100,000 (*see Figure* 2.2). Such data intrigued the nineteenth-century German mathematician Carl Friedrich Gauss, and he discovered that the proportion of primes decreases in a particular way. Among the first 10 (10^1) integers, 1 in 2.5 is prime; among the first 100 (10^2), 1 in 4; among

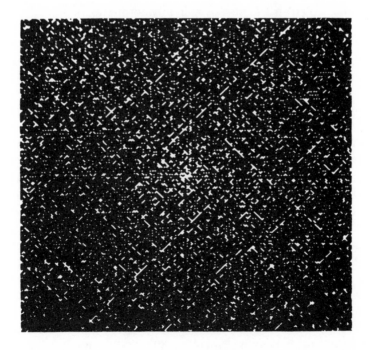

Figure 2.1 A computer-generated grid showing primes (*white dots*) as a spiral of integers from one to about 65,000.

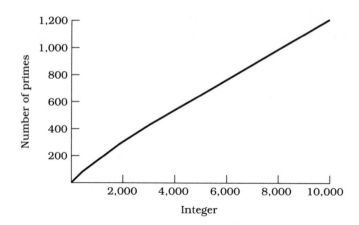

Figure 2.2 Plotting the number of primes (*vertical axis*) less than or equal to a given whole number (*horizontal axis*) illustrates the rate of accumulation of primes.

the first 1,000 (10^3), 1 in 6; among the first 10,000 (10^4), 1 in 8.1; and among the first 100,000 (10^5), 1 in 10.4. Apparently, for numbers up to 10^n, roughly 1 in $2.3n$ is prime.

Gauss conjectured that the number of primes smaller than a given integer x is approximately $x/\log x$, where log stands for the natural logarithm. (The natural logarithm of a number x approximately equals 2.302585 times the base 10 logarithm of x. For example, suppose $x = 1,000 = 10^3$. The base 10 logarithm of 10^3 is simply the exponent, 3, and $\log x = 2.302585 \times 3 = 6.907755$. Hence, the formula predicts that there are $x/\log x = 1,000/6.907755 = 145$ primes up to the value 1,000. The actual number of primes is 168.) A table of logarithms that Gauss used as a lad has a complete statement of the conjecture, written in his hand as a note in the margin.

A few years later, the prominent French mathematician Adrien-Marie Legendre (1752–1833) came up with an equivalent conjecture concerning the average distribution of primes. Gauss himself never published his conjecture, although he mentioned it in 1849 when he wrote to the astronomer Johann Franz Encke (1791–1865). By that time, Gauss could count the number of primes up to 3 million, and his letter to Encke included data to show the striking agreement between the actual number of primes and the number predicted by his formula.

In the decades that followed, several celebrated mathematicians, including Peter Gustav Lejeune Dirichlet (1805–1859), Pafnuty Lvovich Chebyshev (1821–1894), and Georg Friedrich Bernhard Riemann (1826–1866), tackled various aspects of the distribution of primes. It wasn't until 1896, however, that two young mathematicians independently proved the conjecture of Gauss and Legendre. The honor belongs to the French mathematician Jacques Hadamard (1865–1963) and the Belgian mathematician Charles-Jean de la Vallée-Poussin (1866–1962).

The result is now known as the prime number theorem. In effect, it states that the average gap between two consecutive primes near the number x is close to the natural logarithm of x. Thus, when x is close to 100, the natural logarithm of x is approximately 4.6, which means that in this range, roughly every fifth number should be a prime.

The mathematician Tom M. Apostol of the California Institute of Technology has called this theorem one of the most astonishing results in all mathematics. "It describes a simple relation between the primes and the natural logarithm function—which, at first glance, has nothing to do with prime numbers," he commented. "The prime number theorem is important not only because it makes a fundamental, elegant statement about primes and has applications beyond mathematics, but also because much new mathematics was created in the attempts to find a proof."

At the same time, the prime number theorem doesn't eliminate the element of surprise from the realm of primes. It doesn't specify exactly where a particular prime falls. That determination is left to the innumerable observations and experiments that make up the gathering-of-data phase of mathematical research. This data-collection operation now increasingly involves computers, which have extended the list of known primes well beyond the range of hand calculation.

The mining of prime terrain remains far from finished. Indeed, the study of prime numbers is one of the few areas left in mathematics in which the concepts and questions are still simple enough to intrigue both amateur and professional mathematicians. Is there a general formula or a simple rule that generates only prime numbers, even if not all primes? Are there infinitely many twin primes, or consecutive pairs, such as 3 and 5, 41 and 43, and so on? Are there infinitely many primes of the form $2^p - 1$, where p is a prime? Is there an even number greater than 2 that cannot be written as the difference of two primes? No one knows yet. And there are dozens of similar conjectures awaiting proof or refutation.

Sometimes, tantalizing fragments of patterns show up. The sequence 31, 331, 3331, 33331, 333331, 3333331, 33333331 looks promising, because all the numbers in that sequence are primes. Alas, it fails with the next term. The number 333333331 is composite with factors 17 and 19,607,843. Such accidental patterns lead nowhere.

Some simple equations can generate a surprisingly large number of primes. The remarkable formula $x^2 + x + 41$, for example, yields an unbroken sequence of 40 primes, starting at $x = 0$ (*see Figure* 2.3). For values of x up to 10,000, the formula produces primes nearly half the time. Beyond

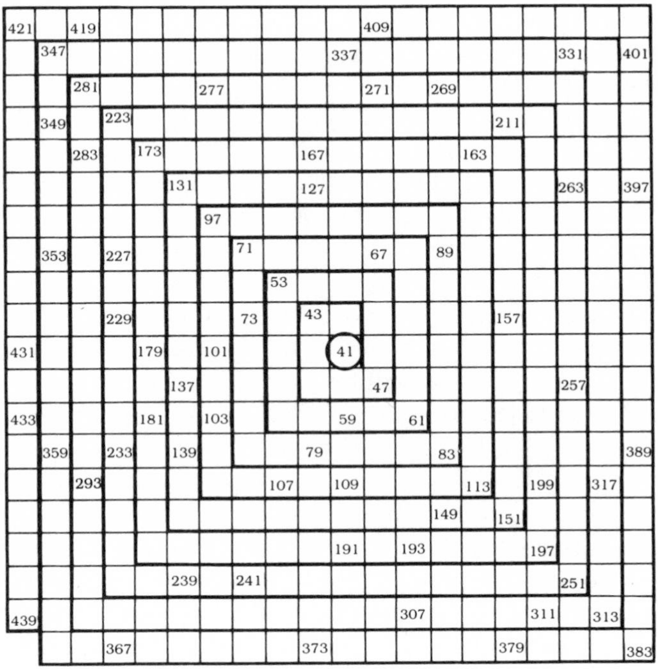

Figure 2.3 The primes between 41 and 439 plotted on a square spiral, beginning with 41 in the center. Values along the diagonal satisfy the formula $x^2 + x + 41$.

10,000, however, the proportion of primes inevitably decreases. Another simple, prime-rich formula, $x^2 + x + 17$, discovered by the prolific eighteenth-century Swiss mathematician Leonhard Euler (1707–1783), generates primes for all values of x from 0 through 15. Of course, one can try other polynomial expressions involving, say, cubes instead of squares. However, mathematicians have proved that no polynomial formula has only prime values, so a search for such a formula to generate all primes is fruitless.

Primes often appear as pairs of consecutive odd integers: 3 and 5; 11 and 13; 41 and 43; 179 and 181; 209,267 and 209,269; and so on. These so-called twins are scattered throughout the list of known prime numbers. Statistical evidence suggests there are infinitely many twin primes, and mathematicians have estimated that in the range close to a given number, n, the average distance between one pair of twin primes and the next is close to the square of the natural logarithm of n. No one, however, has yet proved that there is an infinite number of twin primes. It remains one of the major unsolved problems in number theory. During the last century, numerous mathematicians attacked the problem in a variety of ways, but all failed. The optimists believe that it may take another century of determined assault before anyone succeeds.

At the same time, it's relatively easy to prove that consecutive primes can be as far apart as anyone would want. The sequence of numbers $n! + 2$, $n! + 3$, $n! + 4$ . . . $n! + n$ shows this conjecture must be true. The expression $n!$ (read as n factorial) is less a statement of mathematical excitement than a shorthand way of writing the product of all integers from 1 to n. The number $n!$, where $n = 5$, for example, has a value of $1 \times 2 \times 3 \times 4 \times 5$, or 120. In the general case, $n! + 2$ is evenly divisible by 2, $n! + 3$ is evenly divisible by 3, and so on. Finally, $n! + n$ is evenly divisible by n. Therefore, all numbers in the sequence are composite. The sequence can be made arbitrarily long by picking a sufficiently large number n.

For example, one can readily find a prime gap consisting of three composite numbers. Just looking at a table of primes shows 8, 9, and 10 as the first instance of such a gap. It's also possible to use a formula that guarantees the presence of N consecutive composite numbers: $(N+ 1)! + 2$, $(N + 1)! + 3$, $(N + 1)! + 4$ . . . $(N + 1)! + (N + 1)$. To find a triple $(N = 3)$, one gets the sequence $(3 + 1)! + 2$, $(3 + 1)! + 3$, and $(3 + 1)! + 4$. The numbers are 26, 27, and 28. So, the formula works, but it typically doesn't give the first prime gap of length N. That might occur well before $(N + 1)! + 2$. No one, for example, has yet established where the first prime gap of 1,000 is found, although some progress has been made. Thomas R. Nicely, a mathematician at Lynchburg College in Virginia, has a long-term project aimed at exhaustively searching for prime gaps. By the summer of 1996, his search had reached 3.6×10^{14}, and the largest gap he had identified was 906, which occurred around 2×10^{14}. A gap of 1,000 appears within reach of Nicely's effort. Mathematicians have also obtained research results that suggest strategies for systematically finding consecutive primes separated by a gap of a given length, although not necessarily the first occurrence of such a gap.

In their maddening perversity, prime numbers defy attempts aimed at putting them precisely in their places. To try to prove many conjectures about prime numbers, mathematicians often must devise sneaky stratagems to get at their prey indirectly. They fence in the unruly conjecture, first staking out a wide swath of territory before gradually closing in. The pursuers draw their net tighter, cutting off corners and taking in boundaries while scrupulously checking for loopholes. Finally, if all goes well, the conjecture is trapped.

One such attempt is the continuing work on the Goldbach conjecture, which exemplifies the notion of primes as the building blocks of other numbers. The Prussian mathematician Christian Goldbach (1690–1764) suggested the idea in a letter to Euler in 1742. His claim, expressed in modern terms, is that any even integer greater than 2 can be written as the sum of two primes. The number 32, for example, is the sum of 13 and 19. This proposition is now known as the strong Goldbach conjecture. The weak Goldbach conjecture holds that any odd number greater than 5 can be expressed as the sum of three primes. In both cases, the same prime number may be used more than once in a single sum.

Neither conjecture has yet been proved, but with years of effort, mathematicians have come closer and closer to a final result. In 1937, Ivan M. Vinogradov, a leading number theorist, showed that all "sufficiently large" odd numbers can be expressed as the sum of three primes, which means that if the conjecture is false, there is only a finite number of exceptions. Vinogradov couldn't state explicitly what "sufficiently large" meant, but in 1956, his student K. V. Borodzkin found that $3^{3^{15}}$ works as an upper bound. In 1989, J. R. Chen and Y. Wang lowered that enormous bound to $10^{43,000}$. Although this number is still too large to allow a computer check of all smaller numbers to prove the weak Goldbach conjecture, the trap is closing in, and many mathematicians now regard the conjecture as "essentially proved."

The strong version is somewhat more elusive, but in a famous result reported in 1966, Chen proved that all large enough integers can be expressed as the sum of a prime number and a number that is either prime or the product of two primes. At the same time, in subsidiary research, he also came very close to showing that there are infinitely many twin primes. Computer searches have verified that the strong conjecture holds for every integer up to at least 4×10^{11}. The net tightens.

Because so many important ideas in number theory seem resistant to definitive analysis, experiment and observation play an important role in this field of mathematics. Although their work differs from the experimental research associated with, say, test tubes and noxious chemicals, number theorists, like chemists and other researchers, often collect piles of data before they can begin to extract the principles that neatly account for their observations. Strict reliance on deduction—the hops, steps, and jumps from one theorem or logical truth to another that we usually associate with mathematics—isn't sufficient in number theory.

Despite their successes, some number theorists believe that certain central questions about prime numbers may have no solution. In 1931 a startled mathematical world had to confront and learn to live with such a possibility, not just in number theory but everywhere in mathematics. The logician Kurt Gödel (1906–1978) proved that any collection of axioms, such as those underlying Euclidean geometry or even arithmetic, leads to a mathematical system containing true statements that can be neither proved nor disproved on the basis of those axioms. A theorem, then, can be undecidable. Adding more axioms wouldn't necessarily help, because some theorems would still slip through the inevitable cracks. Furthermore, Gödel's work implies that in some cases, mathematicians wouldn't even be able to decide whether a theorem can or cannot be proved. Uncertainty strikes in mathematics!

Various guesses about prime numbers, perhaps even the strong Goldbach conjecture, may truly turn out to be undecidable. Few mathematical fields other than number theory have so high a proportion of widely believed but unproved propositions. This doesn't mean that the conjectures are all likely to be false, but it implies that for some conjectures, an airtight case might not exist, despite the steady accumulation of persuasive evidence in its favor. Mathematical proofs require more than just overwhelming evidence.

Even so, the quest isn't hopeless. Near misses and proofs that confirm special cases or parts of conjectures hint that eventually techniques may be developed to handle the problems—although it could take centuries. That is far different than saying that a conjecture can never be proved, because a single breakthrough could put such suppositions as the number of twin primes suddenly within reach. Meanwhile, with the use of computers, specific knowledge about prime numbers is growing steadily, although considering the centuries of effort put in by numerous mathematicians, the theory of prime numbers still seems meager.

———— Hunting for Big Primes ————

Success in the search for mammoth primes, which lurk among composite numbers, demands patience, luck, and determination. It also requires careful planning, along with some essential tools from number theory. The quarry are huge numbers such as $2^{44,497} - 1$. The question is whether this 13,395-digit behemoth, just one of an endless supply of candidates, is prime or composite.

Prime hunters do have a systematic way to trap primes and separate them from composites. Inevitably, they use a sieve for such a job. The sieve of Eratosthenes, named for the Greek mathematician who lived in Cyrene in the third century B.C., generates a list of prime numbers by the process of elimination. To find all primes less than, say, 100, the pursuer writes down the integers from 2 to 100. First, 2 is circled, and all multiples of 2 (4, 6, 8, 10, and so on) are struck from the list. That eliminates composite numbers that have 2 as a factor. The next unmarked number is 3. That number is circled, and all multiples of 3 are crossed out. The number 4 is already crossed out, and its multiples have also been eliminated. Five is the next unmarked integer. The procedure continues in this way until only prime numbers are left on the list. In this example, the job finishes with 7 because 8, 9, and 10

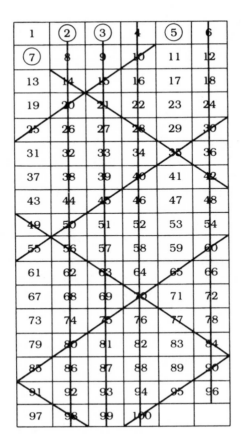

Figure 2.4 The sieve of Eratosthenes catches 25 primes among whole numbers less than 100.

are already gone, and 11 is greater than the square root of 100, the highest number in the table. All multiples of 11 less than 100 would already be crossed out. The sieve ends up trapping 25 primes (*see Figure* 2.4).

The sieve of Eratosthenes represents a systematic way of checking whether a number is prime simply by dividing into the given number all the integers starting with 2 and going up to the square root of the number. If none of the integers divides evenly into the number, the target number is a prime.

In the case of mammoth primes, trial division and the sieve of Eratosthenes prove to be both tedious and hopelessly time-consuming. The real trouble is that trial division provides more information than is required by the question of whether a particular number is prime. The method not only supplies the answer but also gives the factors of any number that happens to be composite. The idea, then, is to find short-cuts that pinpoint primality without finding factors at the same time. If a given number has no small factors, such techniques are bound to be speedier than trial division and its variants.

In the seventeenth century, the jurist and mathematician Pierre de Fermat (1601–1665) provided the basis for primality-testing methods that do not depend on factoring. Today's faster algorithms are all descendants of what is now sometimes called Fermat's little theorem. One version of the theorem states that if p is a prime number and b any whole number, then $b^p - b$ is a multiple of p. In other words, dividing $b^p - b$ by p gives zero as the remainder. For example, if $p = 7$ and $b = 2$, the theorem correctly predicts that 7 divides evenly into $2^7 - 2$, or 126. As in much of his mathematical work, Fermat furnished no proof and no hint of how he came upon the theorem.

Fermat's little theorem provides a handy, foolproof way of showing that a number is not a prime. Suppose that you want to test the number 12. Choosing $b = 2$, $2^{12} - 2 = 4,096 - 2 = 4,094$. Dividing 4,096 by 12 leaves a remainder of 4. Because the remainder is not zero, 12 must be composite. Although calculating the value of 2 (or some other number) to a high power seems more formidable than trial division, mathematical shortcuts via modular arithmetic are available because only the remainder is of interest.

However, finding a remainder of zero doesn't guarantee that a given number is prime. The trouble is that some composite numbers also produce a remainder of zero. These composites are called pseudoprime numbers. The number $2^{341} - 2$, for example, is a multiple of 341, even though 341 is composite, being the product of 11 and 31.

What we need in order to turn the basic Fermat method into an efficient primality test is a way to weed out pseudoprimes. Fortunately, pseudoprimes are rare, although for a given value of b, there are an infinite number of them. Below 10^{10}, for example, there are 455,052,512 primes, but for $b = 2$, only 14,884 pseudoprimes exist over the same range of whole numbers. The composite number 561 is the smallest pseudoprime for all choices of b.

The Fermat test does such a good job of identifying primes that mathematicians and computer scientists have sought to preserve the basic method, merely tacking on techniques for unmasking false primes. Usually, the additional tests provide information to conclude that any divisor of the target number must be among a very small set of integers. If none of the divisors works, the number is a prime. Indeed, the results of such tests on a particular number resemble a mosaic, with information about the number caught up in its fragments. If we look at a sufficiently large piece of the mosaic, we get enough information to decide whether a number is prime or composite.

The pursuit of primes can sometimes lead into novel mathematical territory. Primality testing turns out to be amazingly quick if the hunter doesn't mind making a mistake once in a long while, a result that is good enough for many practical purposes but not for mathematical proof. Such methods represent a trade-off between greater speed and increased uncertainty.

The guessing game hinges on the existence of some numerical property that composites usually possess and that primes never exhibit. Suppose, for example, that for a composite number n, at least three-quarters of the numbers between 1 and n have a particular property that can be checked very quickly. The test simply consists of picking, say, 50 numbers and checking for the appropriate feature. If any have it, n cannot be a prime. If all trial numbers pass the test, n is almost certainly a prime. In that case, the chance of n being composite would be at most $(1/4)^{50}$, or about 10^{-30}. If those odds aren't good enough, trying another 50 numbers takes only seconds, and the chance of error becomes even tinier. Making a sequence of random guesses yields the right answer most of the time, and the longer you keep up the guessing, the better your chances of ending up with a certified prime. Playing the odds can be much more efficient than doggedly following a fixed routine to attain absolute certainty.

Primality testing is also easy when the numbers have a special form. The largest known primes are found among Mersenne numbers, named for Marin Mersenne (1588–1648), a French friar and mathematician and contemporary of Fermat. Mersenne primes have difficulty hiding among the composites because their special structure allows the use of relatively simple tests to determine their primality. They are about as hard to miss as gaily painted circus elephants gamboling in a meadow.

A Mersenne number, M_p, has the form $2^p - 1$, where p is a prime. If M_p itself is a prime, it is called a Mersenne prime. For example, $M_7 = 2^7 - 1 = 127$, which happens to be a prime. In 1644, Mersenne determined that M_p is a prime for $p = 2, 3, 5, 7, 13, 17,$ and 19. M_p is not a prime for 11: $M_{11} = 2^{11} - 1 = 2,047 = 23 \times 89$.

More than 200 years later, François-Édouard-Anatole Lucas (1842–1891) came up with a primality test that turned out to be particularly well suited for testing Mersenne numbers. In 1930, Derrick H. Lehmer (1905–1991) improved the Lucas method to establish the basis of the test still used today. The Lucas-Lehmer test for determining the primality of a given Mersenne number begins with the number 4. Squaring that value, subtracting 2, then dividing by the Mersenne number itself and keeping only the remainder generates a new value, and the procedure is repeated starting with that number. M_p is a prime if, after going through this procedure for $p - 1$ steps, the final remainder is zero. Modular arithmetic strikes again!

For example, if $p = 5$, $M_5 = 2^5 - 1 = 31$. In the first step, the value is 4. In step two, $(4^2 - 2)(\mod 31) = 14$; in step three, $(14^2 - 2)(\mod 31) = 8$; and in step four, $(8^2 - 2)(\mod 31) = 0$. M_5 is a prime!

With the aid of high-speed computers, the Lucas-Lehmer test has turned out to be a relatively quick and easy way to test the primality of Mersenne numbers. In 1978, two California high-school students, Laura Nickel and Curt Landon Noll, logged 440 hours on a large computer to set the record for the largest known prime of that time. Their number, $2^{21,701} - 1$ was the twenty-fifth Mersenne prime discovered and has 6,533 decimal digits.

Until 1996 the pursuit of Mersenne primes and the record for the largest known prime remained essentially a private game for those with access to supercomputers. In many cases, the record primes were found accidentally while a new supercomputer was being put through its paces to make sure the machine was functioning properly. In the spring of 1996, for example, computer scientists at Cray Research unearthed the thirty-fourth Mersenne prime, $2^{1,257,787} - 1$, which comes in at 258,716 decimal digits, in the course of routine testing of a new Cray T94 machine in preparation for delivery to a customer.

In late 1996, however, the record for the largest known prime number went to a computer programmer named Joel Armengaud, who used just a Pentium-powered desktop machine to corner his quarry. It was the thirty-fifth Mersenne prime to be flushed out of hiding: $2^{1,398,269} - 1$ (420,921 decimal digits long). Armengaud was a participant in a remarkable project known as the Great Internet Mersenne Prime Search (GIMPS). Started by George Woltman, a programmer in Florida, it has brought together more than 4,000 volunteers in a systematic effort to check Mersenne numbers to see if they are primes. Because of previous hit-or-miss efforts, not every Mersenne number between the thirty-first and thirty-fifth known numbers had been checked, so it was quite possible that one or more Mersenne primes still lurked in the gaps (*see Figure 2.5*).

Woltman's effort turned computer time wasted on screen savers and other frivolities into mathematical discoveries. "By using a larger number of small computers, we negate the supercomputer's speed advantage," Woltman remarked. "Many other important research projects could use this approach, especially if funding isn't available for months of supercomputer time. It gives the average person a chance to participate in the scientific discoveries of tomorrow."

Meanwhile, the big-prime hunt continues. Where will the next mammoth prime be found? Are there an infinite number of Mersenne primes? No one knows yet.

Value of p	$2^p - 1$	When Proved Prime	Machine Used
2	3	Antiquity	
3	7	Antiquity	
5	31	Antiquity	
7	127	Antiquity	
13	8,191	1456	
17	131,071	1588	
19	524,287	1588	
31	2,147,483,647	1772	
61	19 digits	1883	
89	27 digits	1911	
107	33 digits	1914	
127	39 digits	1876	
521	157 digits	1952	SWAC
607	183 digits	1952	SWAC
1,279	386 digits	1952	SWAC
2,203	664 digits	1952	SWAC
2,281	687 digits	1952	SWAC
3,217	969 digits	1957	BESK
4,253	1,281 digits	1961	IBM-7090
4,423	1,332 digits	1961	IBM-7090
9,689	2,917 digits	1963	ILLIAC-II
9,941	2,993 digits	1963	ILLIAC-II
11,213	3,376 digits	1963	ILLIAC-II
19,937	6,002 digits	1971	IBM 360/91
21,791	6,533 digits	1978	CDC-CYBER-174
23,209	6,987 digits	1979	CDC-CYBER-174
44,497	13,395 digits	1979	CRAY-1
86,243	25,962 digits	1982	CRAY-1
110,503	33,265 digits	1988	NEX SX-2
132,049	39,751 digits	1983	CRAY X-MP
216,091	65,050 digits	1985	CRAY X-MP/24
756,839	227,832 digits	1992	CRAY-2
859,433	258,716 digits	1994	CRAY-C90
1,257,787	378,632 digits	1996	CRAY T-94
1,398,269	420,921 digits	1996	GIMPS
2,976,221	895,932 digits	1997	GIMPS
3,021,377	909,526 digits	1998	GIMPS

Figure 2.5 Since 1952 computers have played a crucial part in the search for Mersenne primes.

Breaking Up Is Hard to Do

Several decades ago, an interest in factoring was the mark of a mathematical eccentric. A small, unheralded group of mathematicians worked quietly, prying open large composite numbers to unlock their prime secrets. They reveled in the pure delight of calculation and in the immense pleasure of devising elegant algorithms to do their work. Like many ardent hunters, they even kept lists of "wanted" and "most wanted" targets (*see Figure 2.6*).

Those lists are still around, but the factoring enthusiasts have been joined by a crowd of computer scientists and applied mathematicians eager to

MOST WANTED

Number	Number of digits in "hard" part
$2^{211} - 1$	60
$2^{251} - 1$	69
$2^{212} + 1$	54
$10^{64} + 1$	55
$10^{67} - 1$	61
$10^{71} - 1$	71
$3^{124} + 1$	58
$3^{128} + 1$	53
$11^{64} + 1$	67
$5^{79} - 1$	55

FACTORIZATIONS

Figure 2.6 The 1983 list of "most wanted" factorizations. Some of these numbers have a small, easy-to-find divisor but that still leaves a "hard" part. All of these numbers were cracked in the space of a year with the use of the quadratic-sieve factoring algorithm.

exercise their hunting skills. That explosion of interest stems from the fact that the security of several important cryptographic systems hinges on the apparent difficulty of factoring large numbers. Armed with advanced algorithms, powerful computers, and specialized calculating machines, the newcomers are rapidly changing the formerly sedate, obscure world of factoring.

As we saw earlier, the most obvious, systematic way to find the factors of a number is by using trial division or the sieve of Eratosthenes. If 2 doesn't divide evenly into the given number, then 3 may, or 5, and so on. Trial division, as fine-tuned by computer scientists, works well for numbers up to about 10 digits in length. There's trouble, however, when the targeted number may be the product of, say, two 100-digit primes. Finding the factors could require on the order of 10^{100} trials—well beyond what a supercomputer could do within a reasonable amount of time.

Mathematicians have long sought alternative methods more efficient than trial division. Fermat, for example, observed that to factor a certain number N, it is helpful to seek out integers x and y such that $x^2 - y^2 = N$. A simple example illustrates the method. If N happens to be 24, choosing $x = 7$ and $y = 5$ meets the equation's requirements. Because $x^2 - y^2$ can be factored algebraically into $(x + y)(x - y)$, N must have the factors $7 + 5 = 12$ and $7 - 5 = 2$. Although 12 can be factored further, this method provides a good start toward a complete factorization of the number 24. The hard part is finding x and y, because random dipping into the pool of available integers is about as futile as locating two needles in a haystack.

What about a situation that requires locating thousands of needles in a haystack that happens to contain millions? That sounds a little easier, especially if the extra work provides a greater chance of success. In fact, mathematicians have created just such a numerical haystack by breaking down the problem of factoring one large number into the problem of factoring thousands of smaller numbers that are relatively easy to manipulate. Because there are millions of such numbers to choose from in any particular factoring problem, the best strategy is to avoid wasting time on any small numbers that prove to be uncooperative. A recalcitrant choice simply loses its place, and another number steps in.

In essence, therefore, it is possible to patch together values for x and y from smaller numbers. In the 1970s, Michael Morrison of the University of California at Los Angeles and John Brillhart of the University of Arizona systematized the patchwork approach to factoring. They established procedures for finding integers that could act as stand-ins for the number to be factored. The hope was that these substitutes would be easier to handle yet would provide the necessary information to flush out the primes of N. Their method made it possible to factor 50-digit numbers with relative ease.

In 1981, Carl Pomerance of the University of Georgia introduced a factoring method he called the quadratic sieve. In effect, he put the Morrison-Brillhart machinery into high gear. Three years later, the new algorithm was used to factor the Mersenne number $2^{251} - 1$. By 1990 the length of numbers that could be factored readily had more than doubled.

A rival of the quadratic sieve appeared in 1988, when John Pollard circulated an account of his number field sieve, based on a somewhat different principle of number theory. Initially, Pollard's method appeared applicable only to a few special numbers. Several mathematicians and computer scientists, however, took it seriously, improved some details of the algorithm, and set about conquering large composites. In particular, they went after the ninth Fermat number, $2^{2^9} + 1$, which was well beyond the reach of the quadratic sieve. In spectacular fashion, Hendrik Lenstra, his brother Arjen, and Mark Manasse succeeded in cracking the number in the spring of 1990. Further refinements greatly improved the number field sieve's performance in factoring not just special numbers but also any that the quadratic sieve could handle. In 1996, the number field sieve became the champion factoring method when it successfully split a 130-digit challenge number in about 15 percent of the time the quadratic sieve would have taken.

Curiously, the quadratic sieve is actually more efficient than the number field sieve for factoring numbers having fewer than 100 digits, and the number field sieve is clearly superior for numbers of 130 or more digits. Where the crossover occurs, however, isn't easy to determine because the performance of an algorithm is highly dependent on fine points in the programming and on the types of computers used.

The steady progress in factoring hasn't yet produced a method that approaches the ease with which numbers can be tested for primality. This may mean that factoring is a genuinely difficult problem for which no truly efficient algorithm can exist. On the other hand, a brand new idea could leap the apparent barrier that current methods appear to face. Mathematicians are still a long way from knowing how difficult factoring really is.

Indeed, two conflicting sentiments lie at the core of factoring theory and practice. On the one hand, mathematicians and computer scientists would like to find new, faster algorithms that would let them factor even larger numbers within a reasonable amount of time. On the other hand, cryptographers who have based the security of their systems on the presumed difficulty of factoring want to be assured that really fast algorithms for factoring large, hard numbers do not exist.

—————— Secrets and Primes ·———— ——

Prime numbers are key figures in a sophisticated mathematical game of hide and seek. The players are a group of mathematicians and computer scientists adept at inventing and solving puzzles. They painstakingly conjure up new methods of writing secret messages while gleefully rooting out the fatal flaws that may lie hidden within rival encryption schemes. Their digital games stretch the boundaries of number theory.

To these puzzlers, deciphering a rudimentary cryptogram such as JXUC-QJXUCQJYSQBJEKKYIJ is little more than a trivial pursuit. In this case, each letter of the original message happens to be replaced by another letter that is a fixed number of places away: for example, an A for a C, a B for a D, and so on. Julius Caesar used this kind of secret code more than 2,000 years ago to hide military information. Modern encryption schemes are much more elaborate and mathematically complex than Caesar's simple cipher. But are they unbreakable? The answer to that question is a major concern for the people entrusted with protecting sensitive data from eavesdroppers, thieves, and spies.

Conventional, modern-day cryptography typically requires a key—a string of numbers—shared by both sender and recipient. The key can then be used in a series of mathematical operations to scramble the digits representing a message. Deciphering the message involves going through the same procedure in reverse, employing the same key.

Cryptosystems can be so secure that the only obvious way to break them is by trying every possible cryptographic key until the solution surfaces. An exhaustive search that requires a computer to work out these trial keys for deciphering the message could easily take years, even with the fastest computers available. Alternatively, a computer could consult a ready-made table listing every possible key for a given cipher, but such a table would probably take up an enormous amount of computer memory. It's no simple matter to discover techniques for breaking these schemes.

However, conventional cryptography has a serious flaw that opens a potential window of vulnerability. The two parties must initially communicate in some way to establish the shared key. Such exchanges are cumbersome and potentially insecure, especially when keys are lengthy and regularly changed to increase security.

In 1976 the computer scientists Martin Hellman and Whitfield Diffie proposed the notion of public-key cryptography to circumvent the key-exchange problem. This method requires the use of a pair of complementary keys instead of a single, shared key. One key, which is publicly available, is used to encrypt information; the other key, known only to the intended recipient, is used to decipher the message. Thus, what the public key does, only the secret key can undo. The security of this type of cryptosystem rests on finding a mathematical procedure to generate two complementary keys such that knowing just the public key and the encryption method is not enough to deduce the private key. The mathematical operation must act like a trap that's much easier to fall into than to escape.

The presumed difficulty of factoring provides such a trapdoor in the RSA public-key cryptosystem, named for its inventors Ronald L. Rivest, Adi Shamir, and Leonard M. Adleman. Using the RSA technique, an individual keen to receive messages would announce a public key consisting of a large number N and an integer r. Anyone wishing to send a message to this individual would transform the message into an integer of length N, dividing the message into blocks if necessary. After each block is raised to some power r and divided by N, the remainder would be sent as the secret message.

The recipient's key is another integer, s, to which no one else is privy. Raising the encrypted message to the power s automatically unscrambles the message. The recipient is safe from illicit eavesdropping, because the only possible way to compute s requires knowing not just N but the prime factors of N as well. Therefore, the recipient has a way of decrypting the message, but everyone else must first factor N before they can get a crack at breaking any transmissions. If N is large enough, that task is practically hopeless. In contrast, because primality testing is quick, the recipient can readily come up with a suitable 200-digit number N by finding two 100-digit primes and multiplying them together.

An example illustrates the RSA scheme's fine points. An official of the Central Security Agency picks any two prime numbers, say, 3 and 11, and multiplies them together to get 33, the value of N. Next, the official goes through a special procedure to come up with the two powers r and s. The procedure's first step is to subtract 1 from each prime to get the numbers 2 and 10, which are multiplied together. The public number, r, is any number between 1 and 20 that is not a factor of 20, which eliminates 2, 4, 5, and 10 from consideration. The number 7 is a suitable choice. A headquarters computer finds a number that when multiplied by 7 then divided by 20 leaves a remainder of 1. Modular arithmetic guarantees that there will always be such a number. The number 3 does the job because $7 \times 3 = 21$, and 21(mod 20) equals 1. Hence, s can be 3.

Headquarters is now ready to send out its spies, all equipped with the public numbers 33 and 7. Agent Alice sends a secret message by converting each letter in the alphabet into a number. The number 2, for example, could represent the letter B. To send B, Alice writes down 2, raises it to the seventh power to get 128, then computes 128(mod 33), probably with the help of a pocket computer. The secret message is transmitted as 29. A headquarters computer decrypts the message by raising 29 to the third power, then calculating 29^3(mod 33). The unscrambled message is 2.

When the RSA cryptosystem was originally proposed, even 50-digit numbers seemed beyond reach of the best available factoring methods and computers. As a challenge to computer scientists, the inventors used their scheme to encode a message, which could be decrypted only if a certain 129-digit number could be broken down into a 64-digit and a 65-digit factor. At that time, Rivest estimated that factoring this number, using the fastest computers and best factoring methods then available, would require considerably more than 40 quadrillion years of computation. However, this prediction didn't take into account some remarkable improvements in factoring methods, as noted earlier in this chapter. Indeed, the existence of the RSA scheme in itself prompted extensive work on factoring.

In 1994 an international team of computer scientists, mathematicians, and other experts accomplished the deed using the quadratic sieve. Launched by Paul Leyland of Oxford University, Michael Graff of Iowa State University, and Derek Atkins of the Massachusetts Institute of Technology, the effort required the use of more than 600 computers scattered throughout the world. Partial results were sent electronically to Atkins, who assembled and passed the calculations on to Arjen Lenstra at Bell Communications Research. In the final step, which by itself consumed 45 hours of computer time, Lenstra used these data and a computer with 16,000 processors to calculate the factors. The entire effort took about eight months.

In the end, the team had their prime factors:

$$114,381,625,757,888,867,669,235,779,976,146,612,010,218,296,721,242,362,$$
$$562,561,842,935,706,935,245,733,897,830,597,123,563,958,705,058,989,075,$$
$$147,599,290,026,879,543,541 = 3,490,529,510,847,650,949,147,849,619,903,$$
$$898,133,417,764,638,493,387,843,990,820,577 \times 32,769,132,993,266,709,549,$$
$$961,988,190,834,461,413,177,642,967,992,942,539, 798,288,533.$$

The decrypted message said: "The magic words are squeamish ossifrage." An ossifrage is a rare, predatory vulture. Its name means "bone breaker."

In one sense, the magnitude of the effort required to factor a 129-digit number demonstrated the strength of the RSA cryptosystem, which typically involves numbers of 155 or more digits. However, steady improvements in factoring methods are likely to force the use of significantly larger numbers in the future to ensure an adequate level of security.

More worrisome, however, are the consequences of new research apparently calling into question any cryptosystem based on factoring. In a startling theoretical result announced at roughly the same time that the 129-digit RSA challenge number was being factored, Peter W. Shor, a mathematician at Bell Labs, proved that factoring, when done on a special type of computer based on quantum mechanical principles, changes from a time-consuming chore to an amazingly quick operation. Although such a quantum computer does not yet exist, Shor's discovery has profound implications for both number theory and computer science.

The theory of quantum mechanics provides a remarkably complete, accurate description of the behavior of atoms, electrons, photons, and other entities on a microscopic scale. Nonetheless, in the everyday world, one doesn't often need to think about this small-scale behavior. For example, a knowledge of quantum mechanics isn't really necessary for designing, manufacturing, or using a screwdriver, even though the hardness and toughness of its metal tip depend on quantum mechanical interactions.

Similarly, it isn't necessary to go back to fundamental quantum theory, as expressed by the Schrödinger equation, to understand enough about how a conventional transistor works to use it appropriately. Simple models involving the motion of electrons and "holes" (corresponding to the absence of electrons) are adequate for determining its overall behavior.

Ordinary, or classical, computers rely on vast arrays of miniature transistors arranged into logic units called gates to perform their calculations. For example, a certain type of gate may switch a 1 to a 0 and vice versa, and another type could take two bits and make the result 0 if both bits are the same and 1 if they are different. Classical computers typically use the presence or absence of certain amounts of electric charge to represent the 1s and 0s of a computer's binary code. Each individual bit must be either 0 or 1, and quantum mechanics doesn't enter into the computations themselves.

In the quantum world, a particle—undisturbed by any attempt to observe it—can be in myriad places at the same time. Thus, a single photon traveling through a crystal simultaneously follows all possible optical paths through the material. In a sense, the photon behaves like an array of waves, and how it emerges from the crystal depends on the manner in which the waves along these different paths reinforce and cancel one another.

Scientists have speculated that computers that operate according to the mathematical rules of quantum mechanics can potentially take advantage of a similar multiplicity of paths to solve certain types of mathematical problems much more quickly than conventional computers. In fact, the notion of quantum computation goes back to 1981, when the noted physicist Richard P. Feynman (1918–1988) remarked that researchers always seem to run into computational difficulties when they try to simulate a system in which quantum mechanics plays a dominant role. The necessary calculations—involving the behavior of atoms, electrons, or photons—invariably require huge amounts of time on a conventional computer. Feynman suggested that a computer based on some sort of quantum logic might circumvent the problem.

In 1985, David Deutsch of the University of Oxford provided the first theoretical description of how a quantum computer might work. Deutsch and a number of collaborators gradually refined these ideas, and they established that a quantum computer could perform certain logic operations that a conventional computer cannot. For example, each bit of a quantum computer, encoded using two different energy levels of a quantum particle, can exist as a combination of the two possible particle states. The 1 and 0 states are said to be entangled. Only when the particle is observed—detected by some instrument—does it settle into one or the other of the two states.

One can program a conventional computer to select—according to the laws of probability—one of several possible computational paths to arrive at an answer. In any specific instance of the calculation, only one of the potential paths is taken, and the choices not made have no influence on the calculation's outcome. In contrast, what makes quantum computation so powerful—and mysterious—is that all potential computational paths are taken simultaneously in a single piece of hardware, in accord with the superposition principle of quantum mechanics. All the possible paths interfere with one another in much the same way that overlapping waves of water can cancel or reinforce each other. Such quantum interference, or superposition, adds a logic element that's missing from classical computation. The trick is to program the computer so that the computational paths that yield undesirable results cancel each other out, whereas the "good" computational paths reinforce each other.

However, although such a machine was potentially more powerful than a conventional computer, Deutsch and others could come up with only highly contrived examples in which that superiority was evident. In 1993, Umesh Vazirani and Ethan Bernstein of the University of California at Berkeley and later Daniel Simon of the University of Montreal established that a significant speedup was possible in certain cases. Inspired by this work, Shor found a way of applying their findings to factoring.

Suppose one wants to find the factors of a particular 100-digit number. With an ordinary computer, one could proceed by dividing the given number by all prime numbers with 50 or fewer digits and looking for any instance in which the remainder is zero. Such a procedure—and alternative, speedier methods of factoring—typically requires an extremely large number of computational steps. It's like looking for a needle in a haystack by checking the straws one by one.

A quantum computer, however, offers the possibility of handling a huge number of computational paths, or states, at the same time. The trick is to express the mathematical problem in a way that takes advantage of this intrinsic multiplicity. Shor devised such an algorithm for factoring on a quantum computer. In effect, his method constructs a haystack in which a hidden needle is transformed into a row of evenly spaced needles. In a quantum computer, such a regularly repeating pattern stands out like a sore digit, and the needle is readily located.

With a quantum computer, once a calculation is set up, computation proceeds simultaneously along many paths according to the specified rules—as long as the computer is left alone to do its work. No one can look inside to check a calculation's progress. Some computational paths reinforce one another, while others cancel each other out, and the computer generates the answer in short order.

Such a procedure runs counter to current thinking in computer science about computing as a step-by-step process. Shor also showed that quantum computation speeds up the calculation of what are known as discrete logarithms, and other researchers have since uncovered additional problems, including searching a database efficiently, where quantum computers could be useful. Building a quantum computer involves surmounting significant technological barriers. Nonetheless, some researchers have already started to produce designs and construct rudimentary quantum logic units that may eventually lead to working models.

————— —— —— — ————— ——

The importance of number theory in cryptography vividly demonstrates how esoteric, playful mathematics, developed with no particular application in mind, can flower into an indispensable element of modern society. Nonetheless, one central difficulty in all this activity is that no one can yet prove mathematically that a given cryptosystem is secure. All that anyone can say is that a lot of people have poked and prodded a given scheme for several years and that nobody has figured out a way to attack it. Current mathematical techniques are good enough to put a fence around a problem to show where security lies. They can demonstrate that the only way in is through a particular gate. How strong is the lock on the gate? Mathematicians still don't have good techniques for answering that question.

Studies of prime numbers also illuminate the role of observation and calculation in the development of mathematical thought. Few people outside of mathematics are aware of the field's empirical aspect. Much of the mathematics encountered by high-school and college students appears carved in stone, passed on unchanged from one generation to another. Yet even the fundamental principles of arithmetic and plane geometry were once the subjects of debate and speculation. It took centuries of constant questioning, brilliant guesses, and steady refinements to build the edifice now known as mathematics, and the structure continues to evolve.

3

Twists
of Space

In his charming book on the wonders of soap bubbles, the British physicist Charles Vernon Boys (1855–1944) wrote: "There is more in a common bubble than those who have only played with them generally imagine." Recent developments in mathematics show that those words, written at the beginning of the twentieth century, remain true. Inspired by the geometry of glistening soap films stretched across wire frames and jostling soap bubbles clustered in unruly masses, mathematicians have ventured ever deeper into a world of exotic geometric shapes.

"Soap films provide a wonderful and accessible physical experiment that leads to many complicated mathematical problems," the mathematician Anthony Tromba of the University of California at Santa Cruz once remarked. "The problem is there. It wasn't invented. Since it does exist, it stands as a challenge to the ingenuity and creativity of the mathematician."

Fairy-Tale Tents

The structures designed by the German architect Frei Otto are as graceful and airy as spider webs. Translucent membranes, supported by steel wire nets, reach out from tall masts. Anchors tie the fringes to the ground. These ethereal forms are also anchored in practical reality. Otto wanted to use as little construction material as possible for enclosures that are easily erected, dismantled, and moved. He drew his models from one of nature's most parsimonious and elegant creations—soap films. They became one of Otto's principal tools in designing exhibition halls, arenas, and stadiums (see Figure 3.1).

The experimenter starts with a plexiglass plate studded with thin rods of various heights. Drooping threads hanging loosely from post to post define rudimentary edges and ridges. Dipping this configuration into a soap solution and gingerly withdrawing it magically transforms the ungainly contraption into a glistening tentlike shape. The soap film, stretching only as far as it must, pulls the threads taut to create a spectacular, scalloped roof.

The soap-film models are then carefully photographed and measured. Solid miniatures are built and tested in wind tunnels to determine the response of these structures to the impact of snow and wind. Finally, plans are drawn up and construction begins, with sheets of synthetic material replacing the soap film and steel cables replacing the threads. With such models, Otto and his collaborators studied forms that had never before been used for buildings. They explored complicated shapes and solved by experiment numerous mathematical problems associated with surfaces and contours.

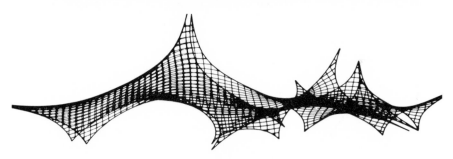

Figure 3.1 Architect Frei Otto used soap-film experiments to design the roofs of several buildings at the 1972 Olympics in Munich. *Top*: Exterior view of the swimming arena. *Bottom*: Drawing of stadium roof.

Many generations of mathematicians have likewise felt the lure of soap films, which serve as excellent examples of the concept of minimal surface. Gently poking the surface of a film stretched across a wire loop—of the type that children use to blow soap bubbles—always increases the film's area. When the disturbance is gone, the soap film springs back to its original shape, again taking on the smallest possible area that spans the loop.

The basic physical principle underlying this behavior is a system's tendency to seek a state of lowest energy. Stretching a soap film increases its surface energy. Work must be done to deform it, in the same way that energy is needed to stretch a rubber band. Whenever it can, a soap film seeks a shape that minimizes its surface energy; and because its surface energy is proportional to its area, it automatically assumes the form of a minimal surface. Consequently, soap films can be used to solve mathematical problems like those that come up in Otto's design work.

Soap-film behavior also suggests ways of solving path-finding problems. Suppose three cities, Altford, Benford, and Camford, are to be connected by a system of highways. Because the cities are all situated on an unnaturally smooth plain and there are no lakes, hills, rivers, or other obstacles in the region, the builders can put their roads anywhere they wish, as long as they keep the cost (and total length) to a minimum. In mathematical terms, the problem comes down to finding a point **P** such that the total length of the straight lines joining **P** to each of the given points, **A**, **B**, and **C**, on a plane has the least possible value. **P** is known as a Steiner point, named after Jakob Steiner (1796–1863) of the University of Berlin, who first devised such problems nearly two centuries ago. For the case in which the three cities form a triangle (that is, they are not all in a straight line), the solution involves finding a point inside the triangle toward which straight roads can be built from each city. Using elementary geometry, one can show that the angles between the lines **AP**, **BP**, and **CP** in this simple network turn out to be 120 degrees (*see Figure* 3.2).

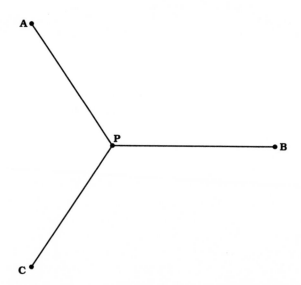

Figure 3.2 A Steiner problem involves finding a point **P** such that the total length of the straight lines joining **P** to each of the three given points has the least possible value.

Fairy-Tale Tents

The Steiner problem can be extended to any number of cities (or points) on a plane and even to situations involving obstacles—the kind that figure prominently in the nightmares of road builders attempting to link cities separated by rugged terrain. In general, given n points, the problem is to find a network of lines that connects all the points and has the shortest total length. For four cities the solution involves locating two Steiner points, but the angles between the roads arriving at these points remain 120 degrees.

The answer can also appear in the shape of a soap film, owing to the fact that the surface tension of a liquid film always acts to minimize the film's area to reach a stable equilibrium. The experimental solution involves building a plastic frame that consists of a certain number of upright pegs, representing cities, sandwiched between two transparent parallel plates. When this model is dipped into and then pulled out of a soap solution, the answer is written in the network of planar soap films linking the pegs (*see* Figure 3.3). The area of the soap film equals the distance between the two plates multiplied by the total length of the liquid edges along the surface of one plate. Because the soap-film system minimizes area and the distance between the plates is constant, the edges viewed on one plate must minimize length.

Obstacle problems come up whenever engineers, planners, or managers must find the optimal route through a thicket of constraints. What trajectory should a spacecraft follow to reach the moon in the shortest time or using the least fuel? What would be the shape of a crystal having the least surface energy for its volume? Whether derived by experiment, worked out in a computer simulation, or extracted using the tools of calculus, soap-film solutions show the way in the branch of mathematics known as the calculus of variations, which deals with finding maxima or minima—the summits or valleys of mathematical landscapes.

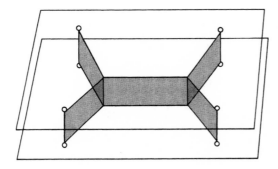

Figure 3.3 The soap-film solution of Steiner's four-cities problem links four points so that the path joining them has the minimum possible length.

Getting to the Surface

A ring dipped into a basin of soapy water comes out spanned by an iridescent film in the form of a thin flat disk. Any other ring-bounded surface, whether barely wrinkled or strongly bulged, would have a larger area. The flat disk is the most miserly surface that a soap film spanning a circle can display. It corresponds to the minimal surface defined by a circle.

When the ring is replaced by a twisted but still closed wire loop, the shape of the corresponding minimal surface is much less obvious. More than a century ago, the Belgian physicist Joseph A. F. Plateau (1801–1883) spent years experimenting with liquid films and pondering the forms that emerged. From his experiments, Plateau concluded that nature always finds a way to mold a soap film so that it spans any loop of wire, no matter how bent.

Replacing the wire by a curve, or contour, and the soap film by a surface turns a specific physical observation into a broad mathematical question: Among all surfaces spanning a given contour find the one of smallest area. That general question has become known as Plateau's problem, and it has long fascinated and perplexed mathematicians.

Soap-film experiments demonstrate how convoluted this question can get. A loop of wire bent so that it looks like the outline of a pair of old-fashioned bulky earphones, for example, can come out displaying a soap film with any one of three different configurations (*see* Figure 3.4). Dipped twice into the same soapy liquid, a given wire frame may show two very different forms. It is even possible that a frame twisted in the right way would emerge displaying a different soap-film form every time it was dipped.

Although demonstrations with soap films and twisted wires reveal important clues about the geometry of minimal surfaces, mathematical questions cannot be settled by experiment. No set of experiments, no matter how extensive, can rule out the possibility that doing the experiment one

Figure 3.4 This example shows that a closed curve can be spanned by three different minimal surfaces.

more time would not lead to some new, unexpected result. A mathematical proof is necessary. That usually entails writing down equations to specify curves and surfaces, then studying the equations to learn about the special features of the shapes.

Given the diversity of forms that a soap film or minimal surface can take on, mathematicians have also developed schemes for putting surfaces into different categories according to various features. One rough but useful classification comes out of a branch of geometry called topology, often described as rubber-sheet or plasticene geometry. In the strange world of topology, where distances have little meaning, a coffee mug and a doughnut are indistinguishable.

Two geometric forms are of the same topological type if one shape can be stretched, squeezed, or twisted until it looks just like the other. Cutting and pasting or tearing are not allowed. A doughnut and a coffee cup are topologically equivalent because it's possible to imagine expanding the coffee cup's handle while shrinking its bowl until all that's left is a fat ring (*see Figure* 3.5). On the other hand, there's no smooth way to transform an ordinary juice glass into a doughnut without punching a hole in the glass. By this reasoning, the surfaces of a sphere, a bowl, and a coin all belong to one topological class, or genus, while a doughnut and a coffee mug belong to another.

In general, the surface of a sphere and any deformed surface, no matter how spiky or twisted, that can be smoothed into a sphere belong to genus 0. A sphere with one attached handle, a doughnut (or torus), and a coffee mug belong to genus 1. A sphere with two handles, a pretzel, and a two-handled soup tureen belong to genus 2 (*see Figure* 3.6). Counting the number of handles determines a surface's genus. Among the three earphone soap films in Figure 3.4, two belong to genus 0 (middle and right) and one to genus 1 (left).

The surfaces considered so far have no boundaries. If we walk around on a sphere, torus, or pretzel, we never come to an edge. On the other hand, the inside of a circle—a disk—is a surface that has the circle as its boundary. A sphere or torus that has been punctured to create a hole in the surface also has a boundary. Such a puncture point is called an end. A sphere with one puncture point can be stretched out to form a disk or even an infinite plane (*see Figure* 3.7, *top*). Anyone who has widened a clay pot's mouth

Figure 3.5 Transforming a coffee mug into a doughnut demonstrates that both objects have the same topology.

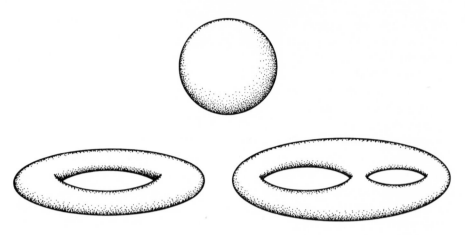

Figure 3.6 Surfaces of genus 0 (sphere), genus 1 (doughnut or torus), and genus 2 (pretzel).

by pulling on its rim has a sense of how such a mathematical deformation works. In the same way, a sphere with two ends can be transformed into a graceful, infinitely extended hourglass form (*see Figure* 3.7, *bottom*). Topological type thus depends on a surface's genus, or the number of handles, and on the number of ends that puncture its skin.

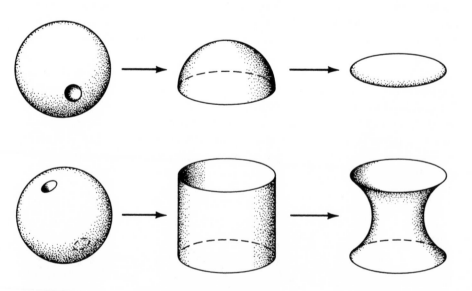

Figure 3.7 *Top*: A sphere with one puncture stretched into a disk. *Bottom*: A sphere with two punctures stretched into a catenoid.

The disk—a sphere with one end—is the simplest possible surface bounded by a closed curve that doesn't intersect itself. The curve itself can be as plain as a circle or as complex as a madly twisted wire, as long as no two of the wire's points touch each other, as they would in a figure eight. Such disk-type minimal surfaces are among the simplest solutions to Plateau's problem.

The mathematical question, then, is whether there exists at least one surface of the disk type having a minimum area among all disk-type surfaces that span a given nonintersecting closed curve. During the nineteenth century, mathematicians solved this particular case of Plateau's problem for many special curves, particularly polygons. In 1928, Jesse Douglas (1897–1965) solved the disk version of Plateau's problem for all simple curves.

Since then, mathematicians have been able to prove that every nonintersecting closed curve is spanned by at least one smooth, area-minimizing surface of finite genus. Nevertheless, the collection of all possible minimal surfaces turns out to be extremely rich. Describing and characterizing these infinite possibilities remains a struggle. Classification by topological type helps, but so do measures of curvature, which play a central role in differential geometry.

Topology focuses on form, whereas differential geometry concentrates on measurable properties, such as surface area and the distance between points. Using the ideas of differential geometry, creatures living on a surface can work out the geometry of their environment not by seeing its shape from the outside but by using instruments to measure its local features.

An intelligent ant, for example, could measure the curvature of its world by pacing off a circle. It would start by establishing a reference point. From that starting point, the ant would step out a certain distance in every direction, marking each endpoint so as to define a circle. The ant would then measure the circle's circumference and compare it with the result of multiplying the circle's diameter by π.

If the ant's measurement equals the calculated answer, the surface is said to have zero curvature. A difference between the measured and calculated circumferences indicates that the surface is curved. The size of the difference, known as the Gaussian curvature, provides an estimate of how curved the surface is. By this measure, a flat sheet of paper has zero curvature, as does the surface of a cylinder, which we can form from a flat piece of paper simply by gluing two opposite edges together. A ball and the inside surface of an ordinary bowl have positive curvature, whereas the surfaces of a cooling tower and a holly leaf typically have negative curvature. A surface of constant curvature is one for which the ant's experiment gives the same result everywhere on the surface. Although both a ball and an egg have positive curvature, only the ball's curvature is constant.

The concept of curvature helps to characterize minimal surfaces. Because the average, or mean, curvature in the vicinity of each point on a

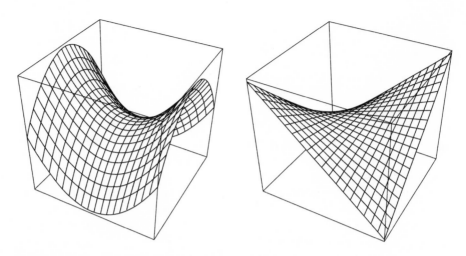

Figure 3.8 A saddle-shaped surface has a mean curvature of zero because the surface curves both downward and upward in the vicinity of every point.

minimal surface must be zero, such a surface is either flat or locally shaped like a saddle (*see Figure* 3.8). A four-legged creature standing anywhere on a saddle surface would find two of its legs sliding down and away from its body and two edging down and closer to its body. That would happen at any spot on a minimal surface, unless the surface happened to be perfectly flat.

Another important measure is that of total curvature. An ant following a circular path turns itself through 360 degrees over the course of one complete circuit. That total amount of turning can be expressed as the number 2π, which represents the total curvature of any circle, regardless of its radius.

Twisted, closed loops of wire have a total curvature that may be greater than the curvature of a simple circle. For example, a wire that loops around twice before its ends join has a total curvature of 4π. As one step in solving the Plateau problem, mathematicians have been able to prove that for contours with a total curvature of less than 4π, only one disk-type minimal surface is bounded by that contour. However, if the total curvature of a curve is even slightly larger than 4π, quite wild and unimaginable things can occur.

Surfaces also have a total curvature. In the case of a sphere, it turns out to be 4π and is again independent of the radius. One method for computing the total curvature for any surface applies a remarkable theorem that connects differential geometry and topology. Established before the turn of the century by G. F. B. Riemann and David Hilbert (1862–1943), the theorem states that any surface is topologically equivalent to a surface of constant curvature. Hence, a rubbery egg and even a malleable cube can always be deformed into a perfectly round sphere. Both objects have a sphere's total curvature—the cube's total curvature is simply concentrated in its edges and corners. No matter how much the object is deformed, its total curvature depends only on its topological type.

The surface of a doughnut—topologically a torus or a sphere with one handle—turns out to have just as much positive curvature as negative curvature. A torus is just a cylinder that's been bent so that its two ends meet, just as a tube of dough can be used to make a bagel. This figure's total curvature is zero. Adding a handle to a torus to make a pretzel decreases the surface's total curvature to -4π. In fact, every additional handle makes the surface's total curvature more negative by 4π.

With the concepts of topological type and curvature, mathematicians can begin to explore minimal surfaces and search for patterns that could lead to theorems about minimal surfaces. Still, progress is slow. Many questions remain unanswered. No one yet knows how to estimate the number of different minimal surfaces that may exist for a given closed curve. Even the much simpler question of how many disk-type minimal surfaces can span a closed curve isn't completely settled.

Meanwhile, topologists have lots of other intriguing surfaces to play with and ponder. One such minimal surface, discovered in 1977 by Bill Meeks during a bout of mathematics at the Gaslight Coffee Shop in Amherst, Massachusetts, contains a Möbius strip. Constructed by gluing together the two ends of a long strip of paper after giving one end a half twist, a Möbius strip has only one side and one edge (*see Figure* 3.9). The

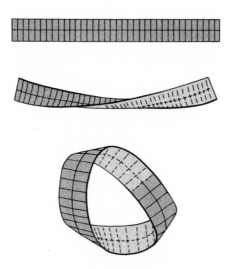

Figure 3.9 A Möbius strip can be constructed by joining the two ends of a long narrow strip of paper after giving the strip a half twist.

minimal surface containing it twists in space, cutting through itself and reaching out to infinity in all directions (*see Color Plate* 3). That strange surface would be enough to dizzy any ant explorer.

———— — — Bites in a Doughnut ———— ——

The disklike soap film clinging to a ring emerging from soapy water is just one small piece of an idealized mathematical object called the plane. In some ways, the plane is like a giant soap film that extends so far over the horizon that the boundary can't be seen. Like a soap film, it is also a minimal surface.

Two rings dipped in a soap solution may each come out spanned by a disklike film. However, a soap film connecting the two rings may also emerge, creating a minimal surface that looks like a pinched cylinder, smoothly taken in at its waist. That hourglass form is the middle piece of another infinite minimal surface called a catenoid (*see Figure* 3.10, *left*). Its two open mouths, one at either end, reach out infinitely into space. Mathematically, a catenoid can be defined in terms of the curve formed by a hanging chain fixed at both ends. Rotating this dangling curve, called a catenary, about a horizontal straight line produces the central section of a catenoid lying on its side.

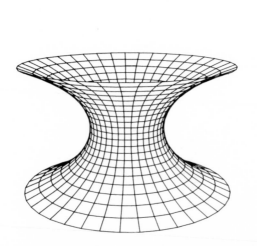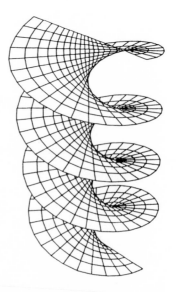

Figure 3.10 Catenoid (*left*) and helicoid (*right*).

A loose coil of wire twisted into the form of a helix supports a spiraling soap film. Extending its ends creates a infinite slide that would whirl any plummeting rider along its surface into a dizzy spin. This infinite minimal surface is called a helicoid (*see Figure* 3.10, *right*).

The plane, catenoid, and helicoid are special minimal surfaces sharing two additional features. Because they are, roughly speaking, without boundaries, they are called complete minimal surfaces; and because none of them folds back and intersects itself, they are said to be embedded minimal surfaces. Until recently, the plane, catenoid, and helicoid were the only known examples of complete, embedded minimal surfaces of finite topology. Topologically, both the plane and helicoid can be modeled on a hollow sphere punctured by a single hole, whereas the catenoid resembles a twice-punctured sphere.

However, topologists had reasons to suspect that perhaps more than three such surfaces actually existed. The trouble was the potential candidates for this minimal-surface hall of fame typically were expressed in complicated equations that masked more than they revealed. This inside information had to be unlocked before the surface's true nature could be seen. Particularly vexing was the problem of determining from the equations whether a surface folds back on itself. Moreover, because these elusive minimal surfaces are infinite in extent, soap-film experiments would be of little use.

The story of the discovery of whole families of new minimal surfaces of this special type begins with a set of equations first written down by Celso Costa, a Brazilian graduate student in mathematics. Costa had managed to prove that his thorny equations represented an infinite minimal surface. Topologically, Costa's surface can be modeled on a three-ended hollow sphere with one handle or, in more tantalizing terms, on a chocolate-covered doughnut from which three separate bites have been taken. The bites puncture the surface and indicate that the surface, when deformed, can extend to infinity in three places.

However, the question remained: Did the surface intersect itself? David Hoffman and Bill Meeks, then at the University of Massachusetts, took up the quest for an answer. They hoped that the surface didn't intersect itself but proving it so was no simple matter. Given a set of equations, there's no single quantity that can be computed to give a yes-or-no answer to the question of self-intersection. All that can be shown is that a particular piece of the surface doesn't intersect another piece, but that's not good enough for an infinite surface, where an infinite number of pieces would have to be compared.

Hoffman's plan was to use a computer to find numerical values for the surface coordinates and then draw pictures of its Swiss-cheese core. These "snapshots" might catch the equations in revealing poses. Mathematical clues already hinted that the surface contains a plane and two catenoids that somehow sprout from the figure's center, but it was hard to see what was happening in the middle.

For assistance with his search, Hoffman turned to James Hoffman (no relation), a graduate student at the university who happened to be developing a new computer-graphics program and language. Together, they spent hours tinkering with the equations and the software necessary to bring the surface to life on a computer screen. David Hoffman knew that if he saw evidence of the surface intersecting itself, the surface would not be embedded and this particular mathematical quest would be over. If there was no visible evidence of an intersection, he could go ahead and try to prove that the surface really was embedded.

The first pictures were full of surprises. The surface appeared to be free of self-intersections, and it seemed to have a high degree of symmetry. It contained two straight lines that met at right angles. "Extended staring," in Hoffman's words, led him to see that the surface consisted of eight identical pieces that fit together to make up the whole figure.

The entranced mathematicians started to explore the surface graphically, rotating it, examining it segment by segment—in effect, clambering about on its contours. Many views later, the true form of this minimal surface began to emerge, and it was strikingly beautiful. The figure had the splendid elegance of a gracefully spinning dancer flinging out her full skirt so that it whirled parallel to the ground. Gentle waves undulated along the skirt's hem. Two holes pierced the skirt's lower surface and joined to form one catenoid that swept upward. Another pair of holes, set at right angles to the first pair, led from the top of the skirt downward into the second catenoid (*see Figure* 3.11 *and Color Plate* 4).

The symmetries revealed in pictures of the figure provided Hoffman and Meeks with just the tips they needed to analyze the equations. That examination, in turn, led to a mathematical proof that the surface was indeed the first complete embedded minimal surface of a finite topology to be found in nearly 200 years. Its predecessors, the catenoid and helicoid, had been discovered in the eighteenth century.

The search didn't end there. Meeks and Hoffman soon demonstrated the existence of an infinite number of surfaces, each one topologically equivalent to a thrice-punctured sphere with a certain number of handles (*see Figure* 3.12). Where there were once only three members of the family, now there were too many to count, and the number continues to grow. A four-ended minimal surface that looks like the original Costa surface cut off at the bottom and then reflected in a mirror was discovered. It started off a second family of complete, embedded minimal surfaces. In this case, the computer was used not only to draw the images but also to search for the correct values of a parameter crucial to the definition of the surface.

The helicoid swept into the minimal-surface limelight in 1992. For centuries, the helicoid was the only known example of a complete embedded minimal surface of finite topology with infinite total curvature. Like the catenoid, this surface has no boundary and doesn't fold back to intersect it-

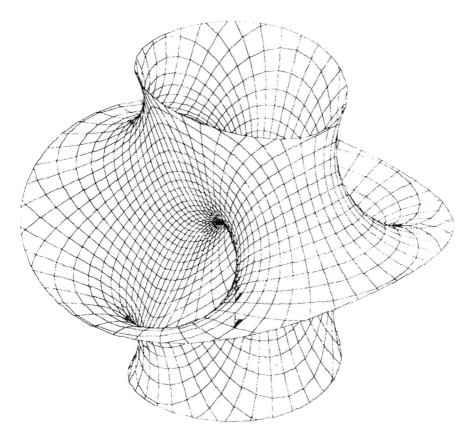

Figure 3.11 The Costa genus-1 minimal surface doesn't intersect itself anywhere. Mathematicians say such a surface is "embedded."

self. At the same time, it's possible to imagine creating a helicoid by carefully deforming and stretching out the surface of a sphere punctured by a hole. The hole's rim becomes the helicoid's spiral edge.

Then, David Hoffman and Fusheng Wei, along with Hermann Karcher of the University of Bonn in Germany, discovered a new minimal surface having the same properties as the helicoid, except that the new surface also incorporated a handle (*see Figure* 3.13). This feature, which looks like a hole in the fundamental helicoid shape, makes the surface topologically equivalent to a punctured sphere equipped with a handle—just like the one that sprouts from a mug. Computer graphics again played an important role in discovering and characterizing this new form.

In general, the computer can be used in situations where physical experiments aren't possible. Computer graphics plays the same role for infinite minimal surfaces that soap films play for surfaces spanning a wire contour.

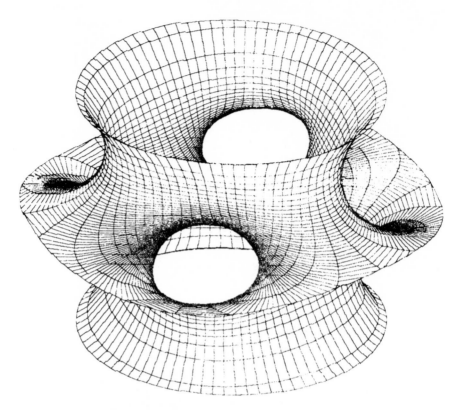

Figure 3.12 Hoffman's genus-3 minimal surface is topologically modeled on a sphere with three handles and three punctures.

It's hard to imagine a wire frame that adequately models surfaces of infinite extent. However, computer experiments are themselves limited in at least one important way. An explicit equation defining a surface is needed to produce a computer picture. Finding such an equation for a surface spanning a given contour is often almost impossible. In contrast, soap-film experiments instantly produce at least one minimal surface for any given wire loop. Nevertheless, when an explicit equation is known, the computer is a useful tool for exploring the defined surface. Because the visual exploration often furnishes clues that can be used later for nailing down the mathematical proof, the machine serves as a guide in the construction of a formal proof.

The new families of minimal surfaces discovered by Hoffman and his coworkers may be of more than mathematical or aesthetic interest. Area-minimizing surfaces often occur in physical and biological systems, espe-

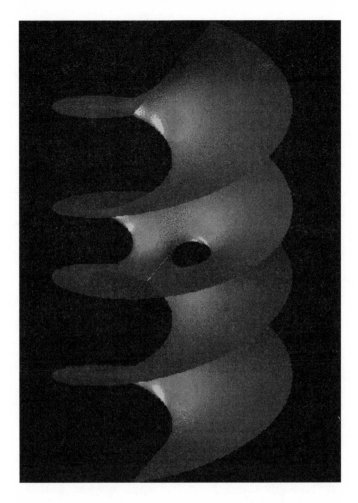

Figure 3.13 A punctured (genus-1) helicoid as a minimal surface.

cially at the boundaries between materials. These newly discovered forms may serve as useful models for understanding the shape of developing embryos, as one biologist suggests, or for the structure of certain polymers. A dental surgeon has noted that such a shape could be used in bone implants for securing false teeth. An implant designed with lots of handles and a least-area surface would minimize contact with bone while still ensuring a strong bond.

Bit by bit, a strangely beautiful new world of minimal surfaces is emerging. Mathematicians probing complex equations are coming up with fixtures for a surreal plumbing-supply store: contorted tubes, helicoids with tunnels, punctured basins. Some forms blossom into bizarre flowers, while others seem ready to fly off the pages of a science fiction novel. Computer graphics is proving to be an indispensable tool that opens up for exploration a hitherto unseen realm of geometric forms.

Bubbling Away

At first glance, a cluster of shimmering soap bubbles filling a basin appears to be nothing more than a haphazard collection of pliant, transparent balls jumbled together. Nonetheless, a delicate balance holds throughout this ephemeral architecture. The popping of even a single bubble triggers a quick readjustment, then all is still again.

A solitary soap bubble's shape is governed mainly by a force known as surface tension, which is uniform over the whole bubble. A soap film enclosing a parcel of air stretches only as far as it must to balance the air pressure inside the bubble—a manifestation of the equilibrium between air pressure and surface tension.

The bubble is spherical because a sphere has the least possible area for the volume that it encloses. Any other shape encompassing the same volume requires a film of larger area. Creating that larger area means stretching the film and therefore proportionately increasing its surface energy. Hence, a soap bubble's perfectly round shape minimizes the bubble's surface energy.

Simple rules appear to govern the way soap bubbles cluster together. It turns out that only three things can happen when soap films meet, even in the largest bubble clusters. First, a smooth sheet of film can separate space locally into two regions, as shown when two bubbles are brought together (see Figure 3.14, top left). Second, if three surfaces meet, as they do when three bubbles come together, they intersect to form a line, and the angle between each pair of sheets is 120 degrees (see Figure 3.14, top right). Third, six surfaces can meet, three at a time, along four edges that come together in a point. That configuration appears when a fourth bubble is placed atop a triangle of three bubbles so that the whole arrangement looks like a tetrahedron. The angle between pairs of edges is always roughly 109.47 degrees, which corresponds to the angle between two crossing diagonals, each drawn from opposite corners of a cube (see Figure 3.14, bottom).

In 1976, Jean Taylor and Fred Almgren (1933–1997) proved that the three basic rules of bubble behavior are a mathematical consequence of area

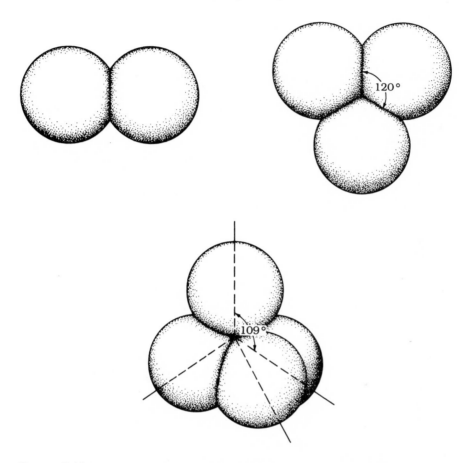

Figure 3.14 Strict rules govern the way two, three, or four soap bubbles come together. Two bubbles form a line; three, a triangle; and four, a tetrahedron.

minimization. For soap bubbles, froths, and films, that principle is related to a physical system's tendency to seek a minimum surface energy at an interface. Such a system will remain in a certain configuration only if it cannot readily shift to one with less energy.

The change in shape can be seen when two bubbles are brought together. Each bubble keeps its spherical shape until the two touch, at which instant the films flow together, eliminating part of the outer surface of each bubble so that the two bubbles share a single, two-sided film. This reduction in film area decreases the total surface area and energy of the original configuration. If the bubbles are of equal size, the interface is flat. If one bubble is larger than the other, the boundary film is a smooth curve that bulges toward the larger bubble.

In the double bubble—a sight familiar to any soap-bubble aficionado, the bubbles share a disk-shaped wall, and this wall meets the individual bubble walls at 120 degrees. Mathematicians call this configuration the standard double bubble. However, this structure isn't the only candidate for the most economical way of packaging a pair of identical volumes. For example, one bubble may ring the other—like an inner tube fitting snugly around a peanut's waist—to form a two-chambered torus bubble (*see Figure* 3.15).

In 1995, Joel Hass of the University of California at Davis and Roger Schlafly of Real Software in Soquel, California, proved that the standard double bubble triumphs over the torus bubble as the two-chambered geometric structure of least surface area. They solved the problem by using a computer to check all the torus bubble configurations, establishing criteria for comparing the surface areas of the various enclosures and writing a program to conduct the search. In the end, they demonstrated that no torus bubble does better than the standard double bubble. Hence, Nature's design also turns out to be the mathematician's answer to the most economical way of enclosing and separating two given volumes of space.

The situation for triple bubbles is much farther from being settled. Moreover, mathematicians haven't even proved that the hexagonal honeycomb is the most efficient way of partitioning a region into equal-area units. Nature probably has the answers already, but mathematicians need to establish them with certainty, and such efforts often entail a lot of toil and trouble.

Clusters of soap bubbles sometimes provide a convenient model of complicated physical processes. A metal's etched surface, for example, typically shows a jumble of grains jammed together. Each grain is a single crystal, made up of atoms in an orderly array. When the metal solidifies, micro-

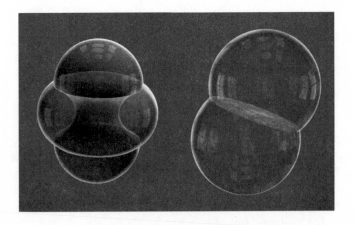

Figure 3.15 A torus bubble (*left*) and a standard double bubble (*right*).

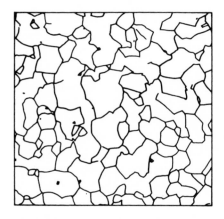 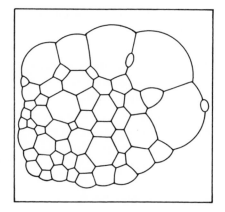

Figure 3.16 *Left*: A deeply etched section of a piece of niobium metal shows a network of grain boundaries. *Right*: A froth of irregular soap bubbles shows a cellular structure analogous to that of metals.

scopic crystals that form within the liquid grow until they bump into their neighbors. The subtle interplay between physical change and chemical force, along with the geometric requirement of filling space, sets the final grain boundaries (*see* Figure 3.16, *left*).

This grain structure often looks a lot like a soap froth (*see* Figure 3.16, *right*) and sometimes behaves like one. Steel, for instance, is a mixture of carbides and iron. When steel is heated up, the grains act much like clustered bubbles, where larger bubbles grow at the expense of smaller ones to create a coarser pattern. A soap-froth model helps metallurgists understand the behavior of metals and suggests ways of manipulating grain structure to get metals with the right properties.

But crystals differ from soap bubbles in crucial ways. Crystal surfaces lack the flexibility of soap films, and this rigidity affects grain boundaries in ways that a soap-bubble model cannot predict. In the 1980s, Jean Taylor of Rutgers University worked with John Cahn, a metallurgist at what was then called the National Bureau of Standards, to get a better idea of the types of boundaries that can form between adjacent crystals. Taylor developed a new kind of mathematics for dealing with what, in effect, are cubic or polyhedral bubbles—forms that have well-defined faces.

As in the case of a soap bubble, surface energy plays an important role in determining the crystal's shape. The cubic form of small crystals of table salt, for example, represents a geometric consequence of minimizing crystalline surface energy. In some directions, the energy required to break apart a crystal may be much lower than in other directions. The plane along which a diamond is most readily cleaved, for example, corresponds to a direction of low surface energy. A diamond crystal's surface energy, then, would be anisotropic, which means that it varies from face to face.

Just as a sphere is the equilibrium shape of a single soap bubble, each anisotropic crystal has an equilibrium shape of its own. This special form, the anisotropic analog of a sphere, is often called a Wulff shape, named for the crystallographer Georg Wulff (1863–1925), who suggested the idea in 1901. Whereas soap bubbles and liquid droplets are all spherical (at least in the absence of gravity and other outside influences), crystals with different chemical compositions may have widely varying surface-energy distributions and, therefore, different Wulff shapes. Depending on the material, a Wulff shape may have the form of a cylinder, a cube, an octahedron, or some other polyhedron.

One of the first steps in studying the interfaces between crystals is to classify the crystal analogs of a soap film on a wire frame. Thus, Plateau's problem of finding minimal surfaces for all possible curves reappears, but in a new, more complicated guise. One useful way to reveal these minimal surfaces is to do the mathematical equivalent of letting a spotlight play over a surface to highlight part of an interface. That means fixing a boundary curve that isolates a section of the surface in order to determine what types of structures have minimal surface energies within that curve. The result is an illustrated catalog of all the interfaces that can occur between a crystal and a surrounding medium, whether solid, liquid, or gas.

Taylor and Cahn initially worked with a Wulff shape in the form of a cube with lopped-off corners (*see* Figure 3.17). The truncated cube, in all its variants, incorporates most common crystal forms. The researchers found 12

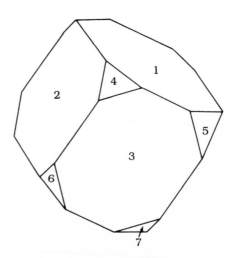

Figure 3.17 This particular Wulff shape appears in the form of a truncated cube.

types of structures that locally minimize surface energy when an interface separates two regions. In contrast, there's only one boundary—a smooth, planar interface—between two neighboring soap bubbles.

Looking like curiously shaped wedges cut from a chunk of cheese, the 12 types of minimal surfaces corresponding to the truncated cube Wulff shape fit into four main categories (*see Figure* 3.18). One category consists of forms that look like the shapes a spotlight would catch if its light happened to fall entirely on one surface of the Wulff shape (a flat disk), on an edge (two planes meeting), or on a corner (three planes coming together). The other three groups feature various saddle shapes.

Several of the interfaces are new to metallurgists. The mathematical proof of their minimality corrects a number of misconceptions that have appeared in metallurgical research papers and other writings. The findings suggest that certain geometries, which metallurgists believed were caused by crystal defects, actually arise naturally in the course of crystal growth. Thus, some features of a metal's crystal structure, once thought to be mysterious or the result of some accident, turn out to be completely reasonable configurations that obey all the known rules.

Taylor's mathematical contribution goes well beyond applications in metallurgy. Her foray into "cubic" bubbles and related anisotropic forms is an extension of centuries of research, inspired by soap-film studies, on isotropic minimal surfaces. The same questions that can be asked about soap films also apply to anisotropic surfaces. Indeed, the study of anisotropic surfaces is a new field in which more is unknown than is known. Basic questions concerning the way surfaces change when the surface energy changes—for example, by shifting the temperature or altering the surrounding atmosphere—remain unanswered. And what about a crystal sitting on a table? Instead of remaining perfectly regular, it sags ever so slightly under the influence of gravity, distorting its geometric form. Beyond all this is the question of how crystals themselves grow—how they attain their perfect geometries in the first place and how they sometimes form the frilly branched shapes typical of snowflakes.

Other mathematicians have recently peered into even more exotic "bubbles" that, unlike spheres, have no counterpart in nature. These strange forms were discovered in 1984, when Henry Wente of the University of Toledo found that the sphere is not the only finite surface having a uniform curvature.

Until Wente's work, no one knew whether a sphere is the only surface with a constant mean curvature. An egg's surface doesn't fit the definition because its surface curvature changes from place to place. An endless cylinder that stretches to infinity in both directions along its axis has a constant mean curvature, but it isn't finite. Indeed, mathematicians had been able to prove that a sphere is the only convex surface of constant mean curvature, but that didn't exclude the possibility of a more convoluted surface fitting

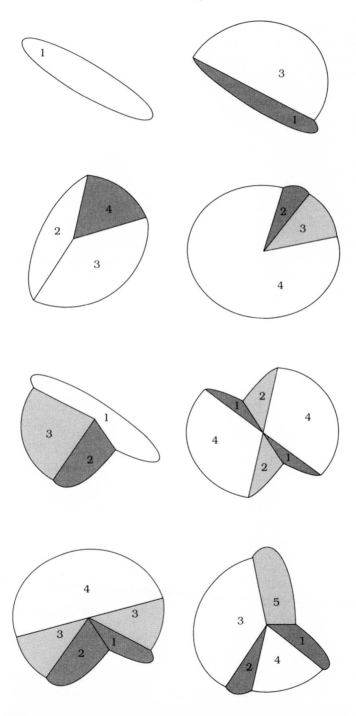

Figure 3.18 Examples from the complete catalog of three-dimensional local structures in equilibrium surfaces of a crystal whose Wulff shape is a truncated cube.

the bill. Mathematicians also knew that any sphere with one or more handles can't have a constant mean curvature unless the surface intersects itself. Circumstantial evidence seemed to go against the possibility that such a surface existed. However, without a mathematical proof one way or the other, the possibility of another constant-curvature form couldn't be ruled out, no matter how great the prestige of those who scoffed at the notion.

Wente's discovery of a strange bubble with constant mean curvature and the proof that accompanied his work now settle the question. Modeled on a torus (that is, a sphere with one handle), his three-lobed toroidal soap bubble bends so sharply that the surface passes through itself in an intricate pattern and hides away its core. The only way to see its full geometry is to slice into the surface and peel it away layer by layer, opening it up like an exotic onion (see Color Plate 5). Computer graphics allows a clear view of the inner works of the three-lobed toroidal soap bubble, one of a family of finite nonspherical surfaces of constant mean curvature now known to exist.

As Plateau demonstrated by experiment and mathematicians have elucidated by analysis and proof, soap films and soap bubbles provide an entrée into a vast realm of endlessly fascinating geometric forms. Even a wire bent into the shape of an overhand knot and dipped into soapy water will emerge with a soap film even though the ends of the wire are not joined together. Knots themselves twist through space in intricate ways, providing further scope for the explorations of mathematicians.

Knotty Problems

In studying knots, mathematicians can get as tangled up in their work as any frustrated individual unraveling snarled fishing line, yarn, or string. In fact, a piece of knotted string with its ends spliced together so that it can't be untied is an excellent physical model for the abstract object with which mathematicians work. A mathematical knot can be thought of as a one-dimensional curve that snakes through three-dimensional space, then catches its own tail to form a loop—like an electric extension cord that's been tangled up and then plugged into itself (see Figure 3.19).

Curiously, "knottedness" is not a property of the curve itself. An imaginary ant crawling along a narrow tunnel within the one-dimensional space of such a curve, even after completing its circuit, would never be able to tell whether the curve is knotted. Knottedness resides in the way the curve sits in three-dimensional space.

Like many aspects of soap-film and soap-bubble geometry, knot theory is part of the mathematical field of topology, in which smoothness, size, and shape can safely be ignored. The only geometric properties that survive are those oblivious to bending, squeezing, twisting, stretching, and other

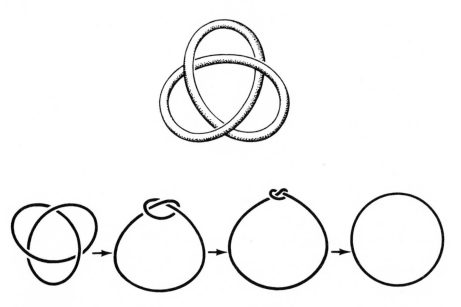

Figure 3.19 Models of knots are constructed by thickening one-dimensional curves slightly so that they seem encased in a solid, flexible three-dimensional tube (*top*), otherwise knots can always be squeezed out (*bottom*).

deformations of space. Complete flexibility is the rule. To topologists, different knots, no matter how twisted or tangled, merely represent different ways of embedding a circle in three-dimensional space.

Mathematicians can ask the same questions about a knotted curve that a boy scout may ask about a knotted rope. What kind of knot is it? Is the curve (or rope) really knotted? Can a second knot undo the first? Is one knot equivalent to another? This last question raises the fundamental problem of knot theory: How can different knots be distinguished?

In general, it's hard to tell if a certain knot tied in a string is the same as a seemingly different one tied in another string (*see Figure* 3.20). One way to solve the puzzle is to try transforming one knot into the other using steps that twist or otherwise deform but don't cut the knot. The trouble with this solution is that it depends on the patience of the person doing the untangling. Spending hours of fruitless labor untwisting and unlooping a knot that still fails to match its companion doesn't prove that the two knots are different. The right combination of moves could somehow have been overlooked.

To attack the problem of classifying and distinguishing knots, mathematicians have adopted a set of rules that make knots more convenient to study. Instead of analyzing three-dimensional knots, they examine two-

1 Crystal Tiles

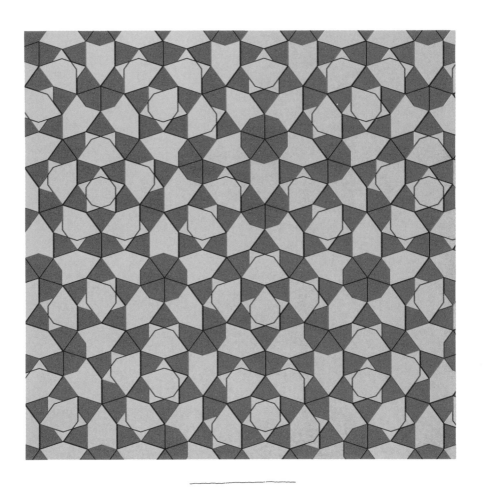

2 Decagon Covers

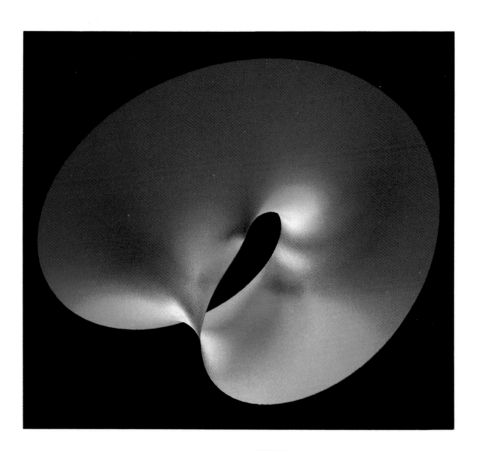

3 Minimal Möbius

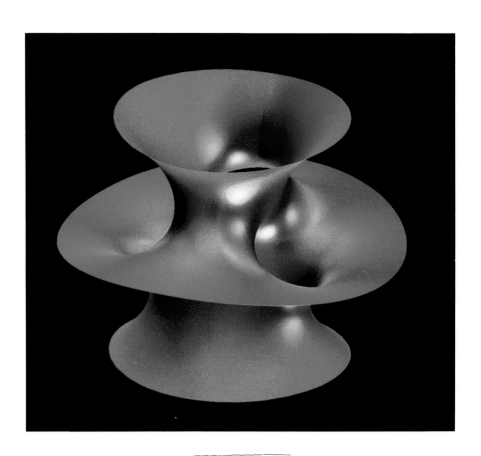

4 Smooth Moves

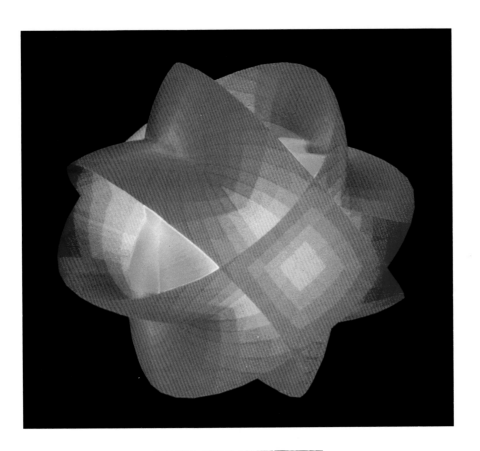

5 Peeling a Psychedelic Onion

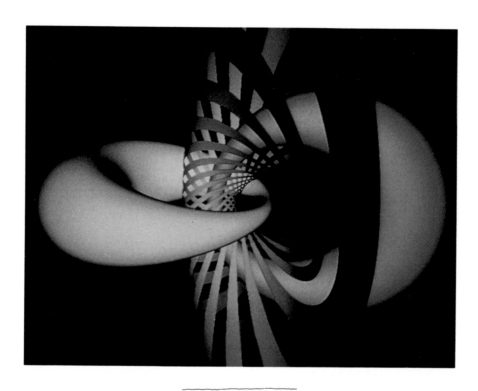

Ó Sliced Doughnuts

7 Mountains of the Mind

8 A Tree Grows in BASIC

Figure 3.20 These two 10-crossing knots look different, but clever manipulation transforms one into the other.

dimensional shadows cast by these knots. Even the most tangled configuration can be shown as a continuous loop whose shadow winds across a flat surface, sometimes crossing over and sometimes crossing under itself. In drawings of knots, tiny breaks in the lines signify underpasses or overpasses, while arrows indicate the direction of travel around a loop (*see Figure* 3.21). Furthermore, just as a suspended wire frame, caught in a breeze

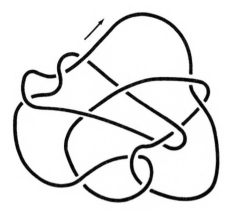

Figure 3.21 By breaking the lines in a drawing of the shadow of a knot and by showing a preferred direction (*arrow*), mathematicians can keep track of a knot's spatial arrangement.

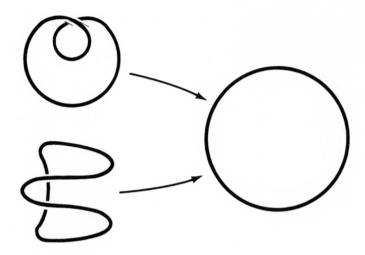

Figure 3.22 An apparent knot can be converted into an unknot.

on a sunny day, casts an ever-changing shadow on the ground, a knot illuminated from different angles can display various projections on a plane. Mathematicians usually work with the simplest projection they can find for a given knot.

A loop without any twists or crossings—in its simplest form, a circle—is called an unknot. Loops that show up in projections with only one or two crossings can always be transformed into unknots (see Figure 3.22). The simplest possible knot is the overhand or trefoil knot, which is really just a circle that winds through itself. In its plainest form, this knot has three crossings. It also comes in two flavors—left-handed and right-handed configurations that are mirror images of each other (see Figure 3.23). No possible deformations of the trefoil knot ever eliminate the crossovers. Only cutting the loop would do the job.

There is just one distinct knot with four crossings and only two with five. The knot club expands rapidly from there to a total of 12,965 distinct knots with 13 or fewer crossings, excluding mirror images. Fourteen is the highest number of crossings for which a complete catalog of knots now exists. Because knots may also be intertwined, like links in a chain, the complexities rapidly multiply.

Sailors, knitters, and other people who regularly work with knots have long had ways to label and classify knots in terms of their physical appearance or characteristics. The mathematical classification of knots, however, goes back only to the late nineteenth century. At that time, the physicist William Thomson (1824–1907), later known as Lord Kelvin, imagined atoms as minute, doughnut-shaped vortices of swirling fluid embedded in a perva-

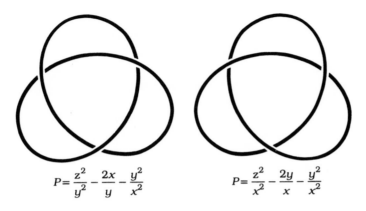

$$P = \frac{z^2}{y^2} - \frac{2x}{y} - \frac{y^2}{x^2}$$

$$P = \frac{z^2}{x^2} - \frac{2y}{x} - \frac{y^2}{x^2}$$

Figure 3.23 A trefoil knot is shown in left-handed and right-handed configurations, with appropriate polynomial labels.

sive, space-filling medium called the ether. To explain what distinguishes one chemical element from another, Thomson turned to knots. He envisioned atoms of different elements as distinctively knotted vortex tubes. Each twisted tube looked like a knotted rope with its two ends joined together in a loop to keep the knot from coming apart.

Intrigued by this idea, Thomson's colleague Peter Guthrie Tait (1831–1901) set out to discover what kinds of knots are possible. This monumental, trial-and-error effort resulted in the first tables of knots, organized according to the minimum number of crossings. However, the procedure was cumbersome and tedious, and the undertaking foundered on the difficulty of determining whether the lists were really complete. The knot theorists had no foolproof method of testing whether two knots are the same or actually curve through space in fundamentally different ways.

To solve the problem of distinguishing among knots, mathematicians tried to develop schemes for labeling them in such a way that two knots having the same label are really equivalent—even though their diagrams may appear quite different—and that two knots with different labels are truly different. In the latter case, the label would be enough to indicate that no amount of twisting, pulling, or pushing would ever transform one knot into the other.

A mathematical label corresponding to any knot property that remains unchanged by deformations is called an invariant. One possible example of such an invariant is the minimum number of crossings found in a drawing of a knot. This number often serves as the basis for organizing lists of knots. In some ways, however, the crossing number is an unsatisfactory invariant. It's not always easy to tell whether a knot has actually been drawn with the minimum number of crossing points. Hence, the number may be difficult to compute. Moreover, this invariant doesn't discriminate very well because many different knots have the same crossing number, and a single number holds little information about the structure of a given knot.

Another approach is to use the arrangement of crossings in a knot diagram to produce a numerical or algebraic expression that serves as a label for the knot. In the 1920s, James W. Alexander (1888–1971) discovered a systematic procedure for generating such a formula from the pattern of over- and under-crossings in a knot diagram. Expressed as a positive or negative power of some variable with integer coefficients, his simple polynomials for characterizing and labeling knots turn out to be remarkably useful and relatively easy to compute, although not ideal.

If two knots have different Alexander polynomials, the knots are definitely not equivalent. For example, the trefoil knot carries the label $t^2 - t + 1$ and the figure-eight knot is $t^2 - 3t + 1$, and both differ from the unknot, whose polynomial is the constant 1. Knots that have the same polynomial, however, aren't necessarily equivalent. The procedure doesn't distinguish, for example, between the granny knot and the square knot even though it is impossible to deform one into the other.

It was several decades before mathematicians fully understood the conceptual base for the calculation of Alexander polynomials and gained a sense of which knot properties the polynomial captures. In Alexander's method, only the direction of the crossings, over or under, and their arrangement with respect to the other crossings make a difference. The formula is the mathematical equivalent of systematically snipping the two strands of the knot at each crossing and refastening the ends so that they are no longer twisted.

For a long time, the Alexander polynomial was one of the few tools topologists had for telling knots apart. In 1984, however, knot theorists were suddenly and unexpectedly thrust into new mathematical territory overrun with novel invariants. The mathematician who triggered the stampede was Vaughan F. R. Jones of the University of California at Berkeley. He found a completely new invariant, another polynomial that does a better job than the Alexander polynomial at distinguishing knots.

Unlike the method for finding an Alexander polynomial, the Jones approach was based on the idea that overpasses and underpasses (or positive and negative crossings) play different roles. His discovery prompted a great deal of excitement in the mathematical community because his polynomial detects the difference between a knot and its mirror image. Jones also caught the mathematics community by surprise because he specializes in an area that has little to do with knot theory. He unexpectedly discovered a connection between knot theory and mathematical techniques that play a role in quantum mechanics.

News of the Jones polynomial set off a wave of mathematical activity. It seemed that just about every mathematician who saw the letter that Jones circulated announcing his discovery and understood its significance caught a glimmer of new possibilities. Mathematicians raced off to look for a general expression that encompasses both the Jones and Alexander polynomials.

The fact that someone reported success in discovering a way to generalize the Jones discovery turned out to be much less of a surprise than that five independent groups of mathematicians arrived at essentially the same result in practically a dead heat. They all found new polynomial invariants that were even more successful than the Jones polynomial in distinguishing between distinct knots. In some mysterious way, these new, improved invariants seemed to delve deeper into the essence of knots.

It was evident that the mathematicians submitting papers had arrived at their results completely independently of each other, although all were inspired by the work of Jones. The whole situation had the potential of a major dispute over priority, but, in the end, it was resolved reasonably amicably. The competing groups eventually agreed that it would be unproductive to try to assess priority. Their answer was to publish one joint paper with six listed authors. A mathematician not directly involved in the discoveries wrote an introduction, and four groups presented summaries of their proofs. The fifth group, a pair of Polish mathematicians, was the victim of slow mail and missed being included in the joint paper. The resulting invariant is now known as the HOMFLY polynomial, after the initials of the last names of the six mathematicians who published the joint paper.

Far from being the end of the excitement, the Jones and HOMFLY polynomials led to the discovery of a host of additional invariants. They also prompted new mathematical questions. Although mathematicians had recipes for computing these new invariants, they had little sense of what features of three-dimensional knots the resulting algebraic expressions encoded.

The Jones polynomial itself remains the key to the puzzle. It apparently encodes many kinds of existing data about knots, but in very strange ways. Many mathematicians are quite amazed that any one polynomial can detect so many different knot properties. Nevertheless, the new invariant has already turned out to be a useful tool. Using it, mathematicians were able to prove something that they had long suspected. A knot in which all the crossings alternate (like the weft in woven fabric) is in its simplest possible form.

It seems clear that all these invariants are part of a still bigger picture that mathematicians barely glimpse. They know that none of the first 12,965 knots has a polynomial that equals 1 (the polynomial of the unknot), but they also know that present theories can't distinguish certain classes of distinct knots. In attempting to unlock the secrets of the new polynomial invariants, mathematicians experiment by drawing lots of pictures and using hours and hours of computer time. They audition knots, looking for qualities that focus attention on what the new invariants reveal and what they hide. To create elaborate structures to test, they carefully paste together relatively simple cases with special properties.

Experience, experiments, and intuition blend in the search for polynomial features that may reflect something observable in a picture of a knot. Like physicists who are trying to make sense of the particles and forces that make up the physical world, knot theorists are looking for something akin to a grand unified theory that would explain all invariants and all knots. They hope eventually to find a complete invariant that distinguishes any two knots.

Knot theory is also making a comeback in the sciences, not as Lord Kelvin's ethereal knotted vortices, but in the twisted, looping, stringlike structures of chemistry and molecular biology. Most famous of these structures is the long, skinny, twisted ladder of deoxyribonucleic acid (DNA) that embodies the genetic code governing life. A single DNA strand, if it had the width of a telephone cord, would be more than a mile long. Outside the cell, the strand looks like a tangled, disorderly heap of spaghetti. In the cell, the pair of helical DNA strands are carefully folded or twisted into a spring-like coil or are joined end to end to form a loop. Two loops may intertwine as links in a chain. Strands may snake around into a knot. Photographs of protein-coated DNA molecules, taken through an electron microscope, clearly show the turns and crossings (*see Figure* 3.24) of such configurations.

Molecular biologists are starting to use knot theory to understand the different conformations that DNA can take on. They can track the sequence of steps in which one structure is gradually transformed into another during the basic, life-supporting processes that take place within cells. Recent advances are helping them see how the enzymes that do the cutting and gluing must perform their functions.

DNA can twist and turn into a number of closed circular forms during biological processes, such as cell division. During such contortions, they can sometimes tie themselves into knots. It turns out that certain enzymes are very good at identifying knots and undoing them. Indeed, the most abundant kinds of enzymes can, almost without error, inspect, recognize, and undo snarls in a knotted DNA circle enormously larger than themselves. Rather than randomly associating DNA ends in the effort to unkink a circle, the enzymes actually simplify the topology by reducing the number of

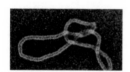 

Figure 3.24 A protein-coated loop of knotted DNA can be untwisted to reveal that the loop is a trefoil knot.

linked molecules. Studies of how enzymes change the topology of DNA strands help researchers determine the details of how they work, and knot theory provides molecular biologists with theorems and models that give them insights into how DNA strands behave in a variety of situations. In this context, one useful concept has been the unknotting number, which is the least number of times required to pass a knot through itself in order to untie it, that is, turn it into an unknot.

Indeed, knot theory has allowed researchers at the University of California at Berkeley to make sense of the confusing array of differently configured DNA strands detected within a cell. It seemed unlikely that a single mechanism could account for the full variety of configurations observed. The researchers had to find out whether two strands of knotted DNA are really the same or different and how one knot related to another—the same questions with which knot theorists have long wrestled. When the researchers arranged the molecules using a mathematical sequence of knots and links, they found a logical pathway. To test their idea, they predicted that a specific six-crossing knot must be the next step, and when they went looking for it, they found the predicted strand (*see Figure* 3.25).

Chemists interested in synthesizing new compounds are also beginning to pay attention to topology and knot theory. In general, they try to create new, unusual molecules by changing the way atoms are connected. In this way, they hope to increase their understanding of how the chemical process of building compounds works. Geometry and especially topology suggest a range of targets for this effort. Inspired by the rigid figures of Euclidean

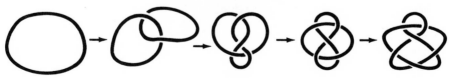

Figure 3.25 By appropriate cutting and splicing, an unknot can be turned into a six-crossing knot (*bottom*). Such a knot was later found among strands of DNA (*top*).

geometry, chemists in the past few years have managed to synthesize such molecules as tetrahedrane and dodecahedrane, in which carbon atoms are connected to form the vertices of a tetrahedron and dodecahedron.

Now that some chemists are beginning to think in terms of the infinitely flexible models of topology rather than the rigid ones of Euclidean geometry, a new set of intriguing targets for synthesis has appeared, including knotted rings and large circles of atoms linked in a chain. Although several syntheses of molecular linked rings, termed catenanes, have been achieved in the past 25 years, no knotted ring has yet been prepared totally by chemical means. The difficulty lies in the unlikelihood of threading a chain of atoms through a loop to form a molecular half hitch. Chemical reactions joining the ends of long strings of atoms are now carried out routinely, but the probability of forming a knotted ring in this way is extremely small.

Hope for creating a tight molecular knot (unlike the looser, naturally generated knots in DNA) has received a boost from a novel technique that relies on the properties of a Möbius strip. Cutting the strip in half lengthwise results not in two separate pieces but in a single strip with four half twists. Other cuts and conformations result in various twisted loops, knotted rings, and separate but linked rings. In 1981, David M. Walba of the University of Colorado managed to create a molecular Möbius strip with a half twist by joining the ends of a ladder-shaped, double-stranded strip of carbon and oxygen atoms.

Walba's work, like that of molecular biologists working with the twists and turns of DNA, has suggested many new mathematical questions. In a crossover that enriches both pursuits, chemists and biologists are drawn into the fascinating mathematical world of knots and links as mathematicians tangle with molecules and chemistry.

———————————————————

"Knot theory for a long time was a little backwater of topology," Joan Birman of Columbia University has remarked. "It's now been recognized as a very deep phenomenon in many areas of mathematics." Aside from links to chemistry and biology, there are intriguing connections between knots and various questions in physics, including the search for a unified theory that incorporates all of the known forces and particles within a coherent framework. To venture into these tangles, we need to look at geometry that stretches from the our everyday three-dimensional world into the difficult-to-visualize realms of four or more dimensions.

4

Shadows from Higher Dimensions

For many explorers of geometry, the doorway to the fourth dimension and beyond has been a slim volume called *Flatland*, written in 1884 by Edwin A. Abbott (1838–1926). The book's central figure and narrator, "A Square," takes tourists into a two-dimensional world where a race of rigid geometric forms live and love, work and play. Like shadows, the denizens of Flatland freely flit about on the surface of their world but lack the power to rise above or sink into it. All Flatland's inhabitants—straight lines, triangles, squares, pentagons, and other regular figures—are trapped in their planar geometry.

"In our own day, there are new reasons to reconsider this book and its fundamental ideas," the mathematician Thomas Banchoff of Brown University wrote in his introduction to the 1991 edition of *Flatland*. "Abbott challenged his readers to imagine trying to understand the nature of phenomena in higher dimensions if all they could see directly were lower-dimensional slices." Radiologists face just such a situation when they build up a picture of the human body from the slices produced by CAT scans or magnetic resonance imaging (*see Chapter* 1). Mathematicians confront a similar task when they attempt to investigate four-dimensional objects by considering three-dimensional "cross sections."

The emergence of computer graphics has brought these strange geometries into play. "We can manipulate objects in four dimensions and see their three-dimensional slices tumbling on the computer screen," Banchoff noted. Theoretical physics draws us even farther afield—into a universe of seven or more spatial dimensions. A century earlier, the original frontispiece of *Flatland* invited readers to contemplate worlds of up to 10 dimensions (*see Figure* 4.1).

Flatland and Beyond

On the surface, Abbott's narrative appears to be simply an entertaining tale and a clever mathematics lesson. From *Flatland*'s beguiling text and quaint drawings, readers can begin to imagine the strictly limited vistas open to those trapped in a low-dimensional realm.

Although their real shapes are two-dimensional, Flatlanders appear to one another as straight lines. Residents of a three-dimensional world see the reason for a Flatlander's restricted view. Observed from above, a coin sitting on a table appears circular. As one's viewpoint shifts closer to the plane of the table, the coin looks more like an oval. At Flatlander level,

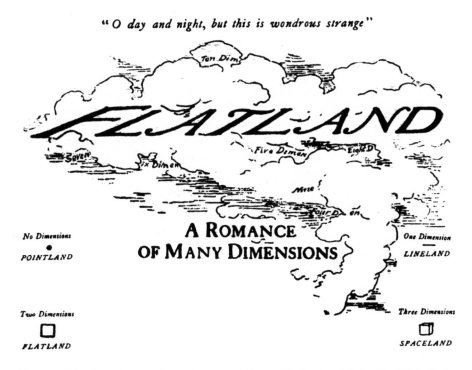

Figure 4.1 The title page from the first edition of Flatland, published in 1884, hints at the book's whimsical nature.

along the table's surface, the oval thins to nothing more than a straight line. Using such examples in Flatland, Abbott neatly introduced key ideas in projective geometry and illustrated a variety of important mathematical concepts.

Flatland is also a sharply delineated satire that reflects widely debated social issues in Victorian Britain. Abbott was a strong advocate of women's rights, and he couldn't resist taking a satirical swipe at his society's attitudes toward women. Flatland women are merely Straight Lines. Lower-class men are Isosceles Triangles; Squares make up the professional class; Nobles are polygons with six or more sides; and Priests, the highest-ranking members, are perfect circles.

"[A] Woman is a needle; being, so to speak, all point, at least at the two extremities," says A Square, the scholarly commentator. "Add to this the power of making herself invisible at will, and you will perceive that a Female, in Flatland, is a creature by no means to be trifled with." Nevertheless, Flatland women also are judged "devoid of brain-power, and have neither reflection, judgment nor forethought, and hardly any memory." In this planar world, men believe that educating women is wasted effort and that communication with women must be in a separate language that contains "irrational and emotional conceptions" not otherwise found in male vocabulary.

When he wrote *Flatland*, Abbott was headmaster at the City of London School, an institution that prepared middle-class boys for professional careers and for places at universities, such as Cambridge. He produced dozens of books, including school textbooks, historical and biblical studies, theological novels, and a well-regarded Shakespearean grammar that strongly influenced the study of the Bard's plays. At first glance, *Flatland* appears out of place within this collection, but a closer look shows that it combines elements of Abbott's broad range of interests, from the nature of miracles to the reform of mathematics education. Abbott was a member of a group of progressive educators who sought changes in the mathematics requirements for university entrance, which at that time included the memorization of lengthy proofs in Euclidean geometry. Abbott's group considered such exercises a waste of time and felt that they narrowed the study of geometry unnecessarily.

Abbott's interest in higher dimensions was also anomalous. Despite evident public curiosity at the time about the concept of a fourth dimension, the mathematics establishment in Great Britain generally refused to admit that higher-dimensional geometries were even conceivable. Conservative mathematicians maintained that such concepts would call into question the very existence and permanence of mathematical truth, as so nobly represented by Euclidean geometry. Abbott challenged such a narrow viewpoint and deliberately called Flatland's university "Wentbridge"—a sly dig at Cambridge.

Flatland also represented one of Abbott's attempts to reconcile scientific and religious ideas and to illuminate the relationship between material proof and religious faith. In one episode, A Square receives a visit from a ghostly sphere, who tries to demonstrate to the bewildered Flatlander the existence of Spaceland and a higher dimension. The visiting sphere argues that he is a "Solid" made up of an infinite number of circles, varying in size from a point to a circle 13 inches across, stacked one on top of the other. In Flatland only one of these circles is visible at any given time. Rising out of Flatland, the sphere looks like a circle that gets smaller and smaller until it finally dwindles to a point, then vanishes altogether (*see* Figure 4.2).

When that demonstration fails to persuade A Square that the sphere is truly three-dimensional, the visitor tries a more mathematical argument. A single point, being just a point, he insists, has only one terminal point. A moving point produces a straight line, which has two terminal points. A straight line moving at right angles to itself sweeps out a square with four terminal points. Those are all conceivable operations to a Flatlander. Inexorable mathematical logic forces the next step. If the numbers 1, 2, and 4 are in a geometric progression, then 8 follows. Lifting a square out of the plane of

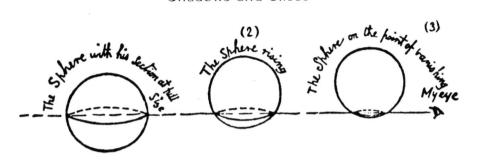

Figure 4.2 To a Flatlander, a sphere passing through Flatland appears as a circle of changing diameter.

Flatland ought to produce something with eight terminal points. Space-landers call it a cube. The argument opens a path to even higher dimensions.

Through mathematical analogy, Abbott sought to show that establishing scientific truth requires a leap of faith and that, conversely, miracles can be explained in terms that don't violate physical laws. Like early scientific theories, miracles could be merely shadows of phenomena beyond everyday experience or intrusions from higher dimensions.

Flatland raises the fundamental question of how to deal with something transcendental, especially when recognizing that one will never be able to grasp its full nature and meaning. Pure mathematicians face such a challenge when they venture into higher dimensions. How do they see multidimensional objects? How do they organize their observations and concepts? How do they communicate their insights? *Flatland* serves as a provocative and informative guidebook for pondering those questions.

Shadows and Slices

Venturing into the fourth dimension can be as dramatic and rewarding as seeing Earth from space for the first time. The view from a higher dimension reveals features and patterns barely noticeable within the lower-dimensional world itself. A truly global picture emerges. Human beings, living in a three-dimensional world, easily recognize the triangles, squares, and other polygons of Flatland, whereas Flatlanders see only lines. In the same way, it takes a view from the fourth or a higher dimension to catch glimpses of the true nature of three-dimensional objects and to see their relationships to one another.

The mathematical journey into the fourth dimension starts with a point—a zero-dimensional object having no length, breadth, or height. A

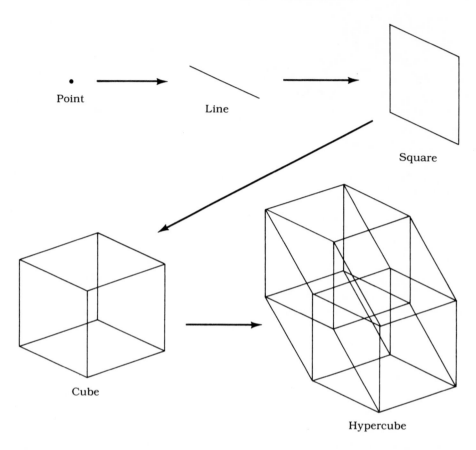

Figure 4.3 A line forms when a point is extended in a particular direction. Shifting the line at right angles to its length generates a square, and moving the square at right angles to its plane produces a cube. Moving the cube at right angles to all three previously defined directions creates a hypercube.

point stretches into a line, which in turn sweeps out a square, which then balloons into a cube. Once a line forms, the figure expands at each succeeding stage in a new direction at right angles to those directions already defined. Each step takes the figure into a higher dimension. Adding a fourth direction at right angles to the other three carries the cube into the fourth dimension (*see Figure* 4.3). The result is a hypercube.

What does a hypercube, or for that matter any other four-dimensional object, look like? One way to keep track of a multidimensional geometric figure is to use a coordinate system to locate the points that make up the object. It takes a single number to specify the position of a point in one-dimensional space, just as a roadside signpost marks the distance from a certain town. Two numbers locate a spot in two dimensions. On Earth's two-dimensional curved surface, latitude and longitude pin down any location. Similarly, x and y coordinates—a pair of numbers—can be used to specify each corner of a square drawn on a sheet of graph paper. In three dimensions, everything is located by three numbers. Hence, a flight over Earth's surface requires an additional, third dimension—the altitude. Logically, four numbers or coordinates must then represent a point in four-dimensional space; five coordinates, a point in five-dimensional space; and so on.

However, most mathematicians need no pictures to step into higher-dimensional mathematical structures. They have long been able to deal with multidimensional problems by treating the concept of dimension more as a mathematical filing cabinet than as a direction in geometric space. Each filing folder in the cabinet represents a dimension in which all the information about one particular variable can be collected and stored. Problem solving and theorem proving are essentially manipulations, according to specific rules, of the numbers and symbols in these folders.

In fact, any set of four numbers, variables, or parameters can be considered either as a four-dimensional entity or as a string of numbers that can be filed and manipulated in certain ways. One of the most famous examples is embedded in the theories of special and general relativity of Albert Einstein (1879–1955). According to Einstein, three-dimensional space and time together make up a four-dimensional continuum in which space and time are intimately interconnected. If we happened to live in a universe of 10 spatial dimensions, time would be the eleventh dimension.

Einstein's space-time is only one of many different types of four-dimensional spaces. A geologist may study past climates by correlating latitude and longitude with the amount of pollen found at various depths in bore holes drilled into soil. The four variables—longitude, latitude, amount, and depth—add up to another kind of four-dimensional space.

Going beyond the fourth dimension is as easy as adding more variables. Some physicists have proposed modifications to Einstein's view of the universe. Their theories introduce seven or more dimensions in the form of miniature hyperbubbles hanging from each point of space-time. Applied mathematicians trying to solve the equations that would allow a business to manufacture and distribute its products most efficiently may deal with thousands of variables, from the cost of raw materials to the distances between cities. The variables are nested in multidimensional spaces impossible to visualize.

What Abbott presented in *Flatland* was the challenge of imagining a four-dimensional geometric space that is homogeneous—one in which every direction looks like every other direction. In other words, from our three-dimensional point of view, we can pick up a four-dimensional box and set it down so that no matter which three of the four dimensions we see, no individual dimension is distinguishable from the others.

Such notions intrigued a few mathematicians early on. In 1827, August Ferdinand Möbius (1790–1868) noticed that the silhouette of a left hand can be turned into the silhouette of a right hand by passing the figure through the third dimension, one higher than the dimension of the figure itself. Anyone can duplicate his mental feat by using a cutout drawing of a hand. Picking up the cutout and flipping it upside down before returning it to a table top transforms the figure into its mirror image. Perhaps, Möbius ruminated, a four-dimensional space would be needed to turn a three-dimensional left-handed glove into a right-handed one.

Later in the nineteenth century, several scholars, including the American mathematician Washington Irving Stringham studied and published drawings of four-dimensional figures. Stringham's sketches in particular and the efforts of C. Howard Hinton (1853–1907), who believed that people could be trained to see a hypercube, prompted a great deal of public interest in the topic. Even the prominent French mathematician Jules Henri Poincaré (1854–1912) discussed the possibility of perceiving the fourth dimension.

Until the computer came along, however, attempts to picture otherworldly objects such as the hypercube had limited success. Human beings had to cope with the same kind of problem that A Square faced when he encountered a sphere in Flatland. As A Square found, it's not easy to get a complete, meaningful view of a higher-dimensional object.

One possible solution is to look at the object's shadow; another is to picture slices through its body. Shadows and slices provide two distinct sets of views that reveal different kinds of information about a geometric object.

Consider the shadow cast by a three-dimensional chair. Although the shadow exhibits distorted lengths, skewed angles, and planes hidden by edges, it still preserves the connections and relationships between the chair's legs, back, and seat. In the same way, the purely mathematical operation of projection, which is akin to casting a shadow, preserves a figure's continuity, although the object's image may be distorted.

Slices through an object provide a sequence of cross sections that preserve the object's angles and lengths, although the relationships among the parts of the object may be ambiguous. For example, cutting along planes parallel to the floor gives a varying picture of a chair. Near the floor, a slice through the legs reveals four disconnected disks. Another slice may capture the square seat, and yet another, the thin rectangle of the back. The mind's eye can reassemble these images into a picture of a chair.

A computer is able to manipulate and plot the four numbers or coordinates that locate each point of a four-dimensional object. Because the screen can display just two or three of the object's four dimensions at one

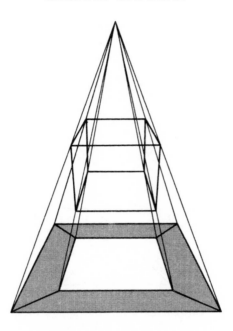

Figure 4.4 A cube with opaque sides and a transparent top and bottom casts a shadow that looks like a square within a square.

time, however, the viewer again sees only shadows or slices of the object. Nevertheless, a computer's speed and flexibility greatly extend the number and types of views of four-dimensional objects at which a mathematician can look.

The shadows cast by an ordinary, three-dimensional cube, transparent except at the edges, gives some clues about what to expect from a hypercube. A cube's two-dimensional shadow can resemble a square within a square (*see* Figure 4.4). Rotating the cube produces a hypnotizing geometric minuet as the shadow's lines lengthen and contract and its squares diminish and expand. So, too, a rotating hypercube springs to life in a movie pieced together from a sequence of computer-generated images. At first, it may look like a square with four corners and four edges. That square turns out to be just one of the six faces of a cube, which appear as parallelograms, sharing a total of 12 edges and eight vertices. Further rotation reveals a starburst of lines—the eight ordinary cubic hyperfaces of a hypercube, which has 24 square faces, 32 edges, and 16 vertices (*see* Figure 4.5).

A similar grand tour of the hypercube seen in perspective is even more dramatic. If a hypercube is illuminated from a point lying outside ordinary three-dimensional space, then its three-dimensional shadow looks like a

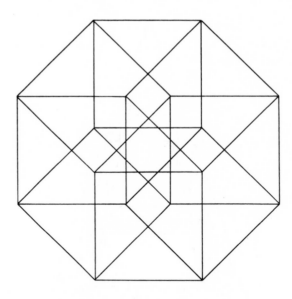

Figure 4.5 This view of a hypercube shows the figure's 16 vertices and 32 edges.

small glass cube floating within a larger glass cube. As the hypercube rotates, the inner cube appears to shift, flatten, and turn inside out (*see Figure* 4.6).

Slices through a cube hint at the hidden riches in store when slices of a hypercube are viewed. Cross sections of a cube may show up as triangles (when a corner is lopped off), as squares and assorted other quadrilaterals, and even as hexagons. In four dimensions, slicing produces stranger shapes—distorted cubes, various prisms, and complex regular and semi-regular polyhedra, often in unexpected arrangements.

Shadows and slices of higher-dimensional objects provide a useful way for mathematicians to visualize these strange forms. Just as experienced musicians can look at a musical score and imagine the sounds, mathemati-

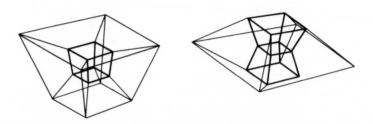

Figure 4.6 Two views of a hypercube rotating in space demonstrate how different the hypercube's three-dimensional shadow may look from different angles.

cians accustomed to working with austere equations and stingy notation can read their ordered symbols and painstakingly construct a mental image of what they have composed. But the mental image isn't always enough. Vivid pictures, like brilliant musical performances in which the sonorous harmonies and subtle patterns written on paper spring to life, bring new richness and beauty to mathematics, which both mathematicians and non-mathematicians can appreciate.

One mathematician who has spent a great deal of time in the shadowy world of four- and higher-dimensional objects is Thomas Banchoff. With computer graphics, he and other research mathematicians have turned the study of geometric phenomena into an observational science. To Banchoff, grand tours of visual images suggest relationships and conjectures that don't arise easily just from equations or lists of data points. In his words: "You can see things before you go ahead and prove them."

Banchoff and his colleagues have created several movies that dramatically show off the startling properties of hypercubes, hyperspheres, and other higher-dimensional forms. Such animations allow mathematicians to explore the curious properties of multidimensional figures. The resulting insights are useful in such mathematical fields as differential geometry in the study of curved surfaces, topology, and data analysis.

A computer-generated sequence of pictures can hold a surprising amount of information. Each tiny dot—called a picture element, or pixel—on a computer screen has a location that can be specified by two coordinates representing two variables. That pixel may also have a color, which could represent a third variable. Animation adds a fourth dimension, usually time. In this way, an astrophysicist can view slices of a complex event, such as the collision of two neutron stars, keeping track of the event in a four-dimensional space of position (two coordinates), density, and time.

The various ways developed in computer graphics for plotting and manipulating four-dimensional data are handy tools in many kinds of data analysis. Often, more than three different types of measurements are necessary to characterize a physical situation. For example, U. S. Navy personnel are interested in uncovering the relationship between the temperature, salinity, biomass, and current strength at particular points in the ocean so that they can track sound waves traveling through water. Computer-generated pictures allow them to search for patterns that help them understand how these various physical factors are related.

Playing with color and relying on various visual illusions to create the impression of height and depth add to the number of dimensions or variables that can be represented in a computer-generated image. That's useful in an age when researchers often have to cope with many variables and billions of pieces of data that otherwise would fill reams of computer paper.

A Global View

From soap bubbles to baseballs and ball bearings, spheres are common-place geometric objects in the three-dimensional world in which we reside. We can play with them, use them, depict them on the computer screen, and write down equations to capture their shape. Mathematically, our everyday sphere is a two-dimensional surface sitting in three-dimensional space; in effect, the skin of a ball. What would a three-dimensional "sphere" sitting in four-dimensional space look like?

To start with, consider the conventional, two-dimensional sphere (or two-sphere, for short). If the sphere happens to have a radius of 1 unit, a remarkably straightforward algebraic formula gives its shape: $x_1^2 + x_2^2 + x_3^2 = 1$, where the coordinates x_1, x_2, and x_3 locate each point of the sphere with respect to three axes at right angles to one another and having an origin coincident with the sphere's center. Just as a globe best represents Earth's features, a three-dimensional view provides the most complete, accurate picture of a sphere. However, it's also possible to represent the surface of a sphere in two dimensions, either on a flat surface or on a computer monitor. Mapmakers do this all the time, having developed a variety of ways to depict continents and oceans on the pages of an atlas.

The standard stereographic projection is one means of achieving the transformation of a globe's surface into a flat map. A simple mathematical formula gives every starting point on the sphere its particular destination on the plane. Suppose the north pole happens to be at the point $(0, 0, 1)$, that is, one unit up the x_3 axis. The projection maps every point on the sphere onto a plane, defined by the x_1 and x_2 axes, which slices through the globe along its equator. According to the formula for a stereographic projection, each point (x_1, x_2, x_3) on the sphere arrives at the point $(x_1/(1 - x_3), x_2/(1 - x_3))$ in the equatorial plane. Thus, a point on the globe's equator stays on the equator, now seen as a circle on the planar map. The point $(3/5, 0, 4/5)$ in the northern hemisphere is sent to the spot $(3, 0)$. In fact, every point in that hemisphere ends up somewhere on the plane outside of the circle marked by the equator. A point, such as $(3/5, 0, -4/5)$, in the southern hemisphere goes to a planar point $(1/3, 0)$, that falls inside the equatorial circle. The south pole ends up at the center $(0, 0)$, while the north pole is banished to infinity. Thus, all the sphere's points, except the north pole, end up somewhere in the plane (see Figure 4.7).

Although the resulting map is a decidedly distorted view of the globe, with all the southern points crowded inside a circle and all the northern points spread out over the rest of the plane, no point on the sphere has been lost and no points overlap. Structures and patterns visible on a globe appear in warped but recognizable forms on the resulting map. By carefully studying what happens to the shape of particular global structures when they are transferred to a flat surface, mathematicians can learn to deduce or

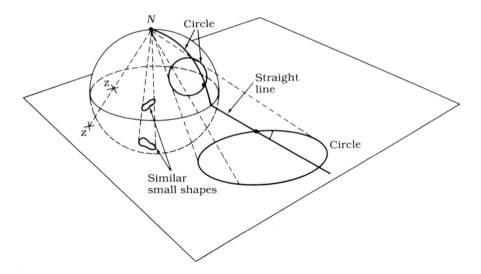

Figure 4.7 The stereographic projection of a sphere onto a plane preserves the shapes of some geometric forms and alters others.

construct global properties merely by examining features on a flat map. For four-dimensional objects, which can't be seen directly, mathematicians have to rely entirely on three-dimensional "maps" to get a sense of their four-dimensional counterparts.

One way to see what's happening under a stereographic projection is to imagine a transparent globe on which circles of latitude have been drawn. If a bright light is fixed at the position of the north pole and the globe's south pole rests on a flat surface, the shadow cast by the circles of latitude is a set of concentric circles. Thus, any circle on the two-sphere reappears as a circle in the plane. The only exceptions are circles that happen to pass through the north pole. They project to a straight line. A set of stripes, all of equal width and parallel to circles of latitude, would project into a bull's-eye of concentric rings on the plane. Each ring gets wider as its distance from the center of the plane, or south pole, increases (*see Figure* 4.8).

The procedure used to create planar images of the two-sphere can be extended to generate images of a hypersphere, or three-sphere, as it would appear in three-dimensional space. The three-sphere in four-dimensional Euclidean space is defined by the equation $x_1^2 + x_2^2 + x_3^2 + x_4^2 = 1$. Compared with the two-sphere, all that's new is the addition of the coordinate x_4.

What does the three-sphere look like? The experience of A Square in Flatland suggests one way to visualize the object. When Flatlanders view a sphere descending from above into their planar world, they see only the part of the sphere that intersects their plane. Their first view is of a point,

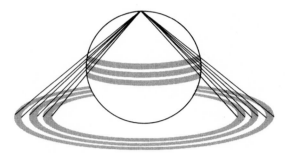

Figure 4.8 Projections onto a plane of bands encircling a sphere.

followed by circles of increasing diameter up to the largest circle, when the sphere's equator passes through Flatland. Then, diminishing circles appear until they dwindle to a point and vanish. By analogy, a human visited by a hypersphere would first see a single spot, or a tiny sphere. That sphere would grow steadily larger to a maximum size, before beginning to shrink, at last disappearing altogether.

Projections, such as the four-space equivalent of the stereographic projection, provide a mathematically more useful glimpse of a hypersphere. One example of such a projection is the Hopf map, named for the Swiss mathematician Heinz Hopf (1894–1971). This map takes points on the three-sphere and systematically finds places for them on the two-sphere. Mathematically, each point (X_1, X_2, X_3, X_4) on the three-sphere becomes the point (x_1, x_2, x_3) on the two-sphere, where

$$x_1 = 2(X_1X_2 + X_3X_4),$$
$$x_2 = 2(X_1X_4 - X_2X_3),$$
$$x_3 = (X_1^2 + X_3^2) - (X_2^2 + X_4^2).$$

Under the Hopf map, every point on the two-sphere represents a circle (called a Hopf circle) on the three-sphere. Reversing the projection by going from the two-sphere to the three-sphere (that is, finding the map's inverse) reveals that a circle of latitude on the two-sphere represents a doughnut-shaped surface, or torus, in the three-sphere. This is enough to build up a picture of the hypersphere. Just as two points (representing the north and south poles) and a series of adjacent bands on a planar map can be used to build up a complete picture of a two-sphere, two linked circles and a family of tori make up a full three-sphere. Hence, a sliced three-sphere can be imagined as a sequence of doughnuts within doughnuts, with each successive doughnut swelling in size as its distance from the center increases (see Figure 4.9).

Figure 4.9 A view of the hypersphere with its image sliced open to expose its inner structure.

By manipulating the computer-generated image of a three-sphere, mathematicians can turn this abstract mathematical object into the star of an animated film. They can slice the object open to get a better view. They can watch the object move, linking strings of images to reveal coherent patterns, such as the smoothly twisted tori that swell up and sweep past one another when a hypersphere rotates (*see Color Plate* 6).

Pictures alone, however, don't tell the whole story. The Hopf map itself was the first example of a special type of projection of points from one sphere to another sphere of lower dimension. Its existence inspired further mathematical research in various geometric topics, including the sorting out of several types of three-dimensional surfaces in four-dimensional space.

Shapes such as the hypersphere show up in applied mathematics and physics. Different sets of mathematical equations, which may be useful for describing physical processes as diverse as fluid flow and crystal growth, generate different curves and surfaces in space. The changing behavior of complex systems over time turns out to correspond neatly to motion on the curving shapes of certain surfaces. In many instances, understanding the solutions of particular equations can be turned into questions of geometry. Although some equation solutions show up as collections of points on a sphere or a torus, others may reside on more exotic objects, such as a one-sided, one-edged Möbius strip or a three-dimensional tube called a Klein bottle, whose inside surface loops back on itself to merge with the outside.

The hypersphere, for one, makes an appearance in the analysis of motion in phase space, in which each dimension represents one of the variables in the equation used to model a particular system. For a mechanical system, the variables may be just the positions and velocities of each particle in the ensemble. Beginning at a point representing the initial values of all the variables, the equation generates a trajectory that winds through phase space. The location of a point on the trajectory at any time contains all the information needed to describe the state of the system—that is, the values of all the variables—at that particular time. Although the motion itself isn't directly visible, such phase-space portraits provide a useful collection of information about the motion.

For example, in the movement of a pair of oscillators, such as two windshield wipers, it takes one variable, the angle, to define each wiper's position and another variable to define each wiper's velocity. Together, the positions and velocities of both wipers trace out a path on the surface of a hypersphere in four-dimensional space. A similar analysis applies to a pendulum bob fixed to a stiff rod and mounted so that the bob is physically free to trace out a path on the surface of a sphere. It takes two coordinates to define the bob's position and two to specify how quickly it is moving. Tracking the pendulum's path through phase space—that is, plotting its position and velocity at different times—gives a graphical history of the pendulum's behavior. Four-dimensional space, however, has so much room that this curving, four-dimensional path could wander almost anywhere. Luckily, physical constraints—the laws of conservation of energy and momentum—keep the system within strictly defined bounds. The four-dimensional phase-space curves are actually confined to a three-dimensional surface.

In normal situations, these curves, as seen in four-dimensional phase space, fall on the surface of a doughnut, or torus. Each torus represents a certain combination of energy and momentum, equivalent to the initial push given to a pendulum to get it started. In a hypersphere picture, the Hopf tori represent the constant energy-momentum surfaces, and the Hopf circles on them are curves representing the changing positions and velocities of each pendulum (*see Figure* 4.10). These curves are essentially solutions of differential equations, called Hamilton's equations, which define how the motion changes with time. We take a closer look at the behavior of differential equations in Chapter 6.

Computer graphics provides a way for researchers to study what the solution curves look like for various physical systems and sets of initial condi-

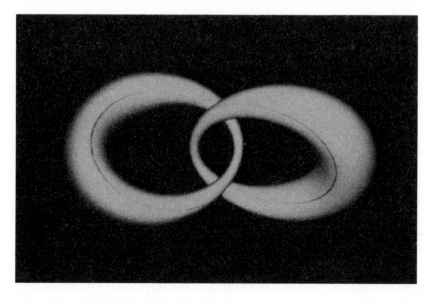

Figure 4.10 Two linked toroidal energy-momentum surfaces for a pair of linear harmonic oscillators.

tions. These pictures greatly enrich the study of dynamical systems—the way things change in time and space.

The Classification Game

Like botanists, mathematicians have a penchant for classification. Botanists, however, can never be sure that every plant species belonging to a particular family has been collected and identified. In contrast, mathematicians can sometimes prove that a classification scheme for a given set of mathematical objects is complete. They can formulate an ironclad guarantee essentially stating that no one will ever discover an object belonging to the set that doesn't fit one of the mandated categories. What keeps the game interesting, however, is that such proofs are often elusive. To mathematicians, classification efforts serve as stringent tests of the power of their mathematical methods.

One of the greatest classification quests in mathematics is the search for a complete categorization of geometric shapes. The task was relatively easy centuries ago, in the days when the recognized geometric forms encompassed familiar, well-defined, rigid figures: lines and circles, triangles and cubes, cones and cylinders, assorted polygons and polyhedra. Then topology reared its rubbery head, and flexibility became the rule. On top of that, the spread of geometry into higher dimensions left mathematicians with an unruly zoo of geometric forms. Attempts to classify these myriad shapes tended to lag behind discoveries of new ones. It was as if a botanist, not yet finished with Earth, were faced with a rapid succession of new planets of exotic species, all defying easy identification.

One of the major triumphs of nineteenth-century mathematics, and of topology in particular, was the complete classification of two-dimensional surfaces, which can be pictured as the surfaces of three-dimensional solid objects. As outlined in Chapter 3, all surfaces, if they are treated like plasticene sheets, can be bent, stretched, curved, or deformed—as long as they are not torn or glued—until they match a set of basic forms. Every conceivable surface, from the skin of a doughnut to the wrinkled exterior of a prune, is equivalent to a sphere with a certain number of attached handles and a certain number of punctures that penetrate its skin. Because the number of handles (known as the genus) of a surface never changes no matter how the surface is deformed, this quantity stands as a way to classify surfaces. Hence, the surface of a coffee mug and the skin of a doughnut, having the same genus, fit in the same category—a one-handled sphere.

When topologists started to explore the world of geometric forms in higher dimensions, they found that neither the intuition nor the vocabulary of ordinary geometry was sufficient to describe and classify the new forms they discovered. They introduced "manifold" as a general term for describing a certain common type of higher-dimensional geometric object. In everyday parlance, a manifold is a pipe or chamber bristling with subsidiary tubes. Its mathematical manifestation encompasses surfaces that locally appear flat, or Euclidean, but on a larger scale may bend and twist into exotic and intricate forms.

A circle, although it curves through two dimensions, is an example of a one-dimensional manifold, or one-manifold. A closeup view reveals that any small segment of the circle is practically indistinguishable from a straight line. Similarly, a sphere's two-dimensional surface, even though it curves through three dimensions, is an example of a two-manifold. Seen locally, the surface appears flat. Roughly speaking, Earth's surface is an example of such a manifold. To a gardener marking off a square plot in a field, Earth's curvature is barely noticeable. The gardener doesn't have to worry about opposite borders that aren't quite parallel or corners that aren't quite at right angles.

The theory of manifolds arose in the nineteenth century out of a need to understand solutions of equations expressing relationships between two or more variables. For instance, all the possible solutions of a two-variable equation can be plotted as a set of points in the plane. Each point represents a pair of values that satisfies the equation. Those points typically fall on a curve. Similarly, the solutions of an equation that has three variables can be plotted as a two-dimensional surface in three-dimensional space. The points could, for example, fall on the surface of a sphere or some other two-manifold. For an equation with more than three variables, the solutions may look like a multidimensional manifold embedded in a space one dimension higher than the manifold itself.

It takes considerable mental agility to picture even a simple three-manifold, such as a three-dimensional sphere, or three-sphere, or something more complicated, such as the space around a loosely knotted string. Three-manifolds come in a much greater variety than ordinary surfaces. Just think of all possible knots and then all possible knots intertwined with other knots. The potential complications are endless. And knots are the models for only a few of the simpler kinds of three-manifolds. Indeed, the mathematics of three-manifolds turns out to be exceedingly intricate and more difficult than the mathematics of manifolds in any other dimension.

In dealing with a three-manifold, a human being in three-space would have the same trouble that a nearsighted ant crawling on the surface of a rather large doughnut would have in telling whether the doughnut has a hole. The ant can't stand outside the surface and see the hole, and it can't step off the surface to fall into it. The hole can be seen clearly only from a higher dimension. Similarly, a human explorer wandering around on a three-manifold wouldn't see much out of the ordinary. It takes a vantage point in four-space to detect holes and other features of a three-manifold.

However, we can try some indirect approaches to the hole-detection problem. An inquisitive ant, wondering whether its surface is a sphere, torus, or some other exotic form, could detect a hole by taking circular tours around the surface. The ant carefully notes the properties of these closed loops as they shrink in size. Sometimes the ant's path forms a closed loop that happens to thread its way through a hole. In this case, the ant would find that its circular path can't be shrunk beyond a certain point; it gets snagged on the hole. Loops that don't thread a hole, however, can be shrunk. If an ant finds that all closed loops on the surface can be shrunk, it knows it's on a sphere.

The ant's experiment can be turned into a useful mathematical procedure for detecting holes. This procedure forms the basis of an algebraic method, developed by Poincaré at the turn of the century, to detect and study the pattern of holes in a topological surface. His strategy, like the ant's, depends on the shrinking properties of loops on a given surface.

All closed paths on a sphere can be continuously shrunk to a single point. That sphere is thus said to be simply connected. In contrast, there are fundamentally different types of paths on a torus, only some of which can be shrunk to a point (see Figure 4.11). Those winding through the center hole differ from those winding around the center hole. And both are different from

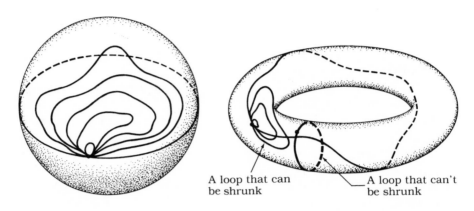

A loop that can be shrunk

A loop that can't be shrunk

Figure 4.11 Loops on a sphere and a torus. Some loops can be shrunk to a point but others cannot.

the cyclic paths that loop through the center as they go around the torus. Every surface other than a sphere contains some unshrinkable loops.

The question is whether the same shrinking loop argument applies to three-spheres (the three-dimensional surfaces of four-dimensional spheres). Poincaré conjectured that applying the loop test to three-manifolds would succeed in distinguishing three-spheres from all other three-manifolds. No fake spheres, which are not simply connected, would pass the test and slip through.

The Poincaré conjecture, still unproved despite numerous heroic efforts over the past few decades, ranks as one of the most challenging and baffling of unsolved problems in mathematics. Whole fields of mathematics have been developed in the course of testing, on a case-by-case basis, not only the conjecture itself but also the extensions of the conjecture to manifolds in higher dimensions.

The central problem in proving the Poincaré conjecture is that, unlike the two-manifold case, no one knows a classification scheme for three-manifolds. There is no list of all possible surfaces that can be checked one by one to make sure that all nonspheres contain unshrinkable loops. Even worse, attempts to classify three-manifolds seem to depend on knowing that the Poincaré conjecture is true. The only way out is to develop totally new methods that bypass the unsolved classification problem.

Ironically, the first major advances toward proving the Poincaré conjecture occurred not for three-manifolds but for higher-dimensional manifolds. One important tool for the study of higher-dimensional manifolds, developed in the early 1960s, was the technique of surgery, that is, systematically studying the effect of removing certain pieces of a manifold, then gluing

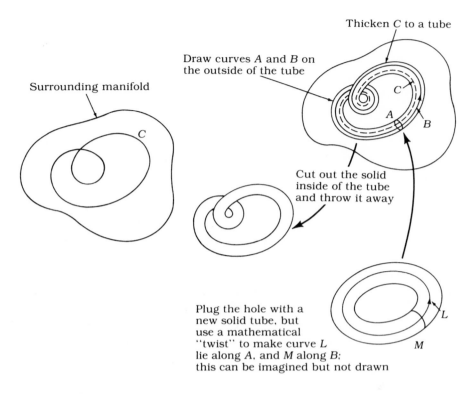

Figure 4.12 An example of topological surgery—removing a tube and then gluing a new one back in with a "twist."

them back with a specific twist (*see Figure* 4.12). In 1961, Stephen Smale developed and used just such a surgical method to break manifolds into digestible, bite-size pieces that could be easily moved around. His work led to a far-reaching theorem, one of whose consequences was the truth of the Poincaré conjecture for five- and higher-dimensional manifolds.

These higher-dimensional manifolds turn out to be easier to classify than three- or four-manifolds, but even spheres, the lowliest of higher-dimensional manifolds, sometimes display unexpectedly bizarre properties. For example, in 1959, John Milnor, doing a kind of multidimensional plumbing—cutting holes here and there, connecting them by various tubes, then deforming the result into a new kind of sphere—stunned the mathematical world by showing that the seven-dimensional sphere can be made into a smooth manifold in 28 different ways. Milnor's result means that Poincaré's scheme of representing the solutions of differential equations on manifolds would, in dimension seven, encounter 28 different versions of the sphere.

After a spurt in the 1960s, efforts aimed at proving Poincaré's conjecture stalled. Two cases remained: three-manifolds and four-manifolds. That left mathematicians with the feeling that something unusual happens in spaces of dimensions four and five. In both, the number of dimensions is high enough to allow complicated behavior but too small to leave sufficient room for mathematical maneuvering of the standard kind to eliminate those complications.

The central problem in classifying four-manifolds involves the technical distinction between topological manifolds and smooth, or differentiable, manifolds. A ball, for example, has a smooth continuous surface. A closed empty box has a continuous surface, but because it has sharp edges and corners, its surface isn't smooth. The difference is crucial because topologists have ways of mathematically smoothing any sharp edges, creases, and jagged features in dimensions one, two, and three. For example, the surface of a box can be smoothed out in a reasonable way that transforms it into a sphere. In other words, there's no difference in these dimensions between topological manifolds (the more general category) and smooth manifolds. In five dimensions and higher, manifolds come in both the smooth and crinkly varieties, and mathematicians have a good understanding of when and how the different types occur. In four dimensions, the distinction between smooth and crinkly manifolds is much more complicated and difficult to sort out.

A similar distinction applies to transformations designed to test whether manifolds belong to the same class or fit into different categories. Like a lump of pizza dough, a topological manifold is free to be kneaded and distorted. In general, two such manifolds can be regarded as equivalent if one can be transformed into the other without tearing. Such transformations may involve a smooth transition or follow a crinkly course. Calling specifically for a smooth transition is a more stringent condition than simply showing that two manifolds are topologically equivalent, somewhat similar to what subspecies a particular flower belongs.

The first major step in classifying four-manifolds was a proof that certain types could be identified on the basis of algebraic invariants—computed formulas that distinguish one type of manifold from another—called quadratic forms. Michael Freedman of the University of California at San Diego took seven years to crack the problem. His 1981 proof showed that such manifolds can be constructed from simple building blocks and classified entirely on the basis of their quadratic forms. Freedman's remarkable work not only proved the Poincaré conjecture for four-manifolds but also unearthed many new examples of four-manifolds.

Freedman's results left just the original Poincaré conjecture unsolved—the close-to-home case of three-manifolds. Mathematical surgery suggests why the three-manifold case is so complex. Surgery makes it possible to

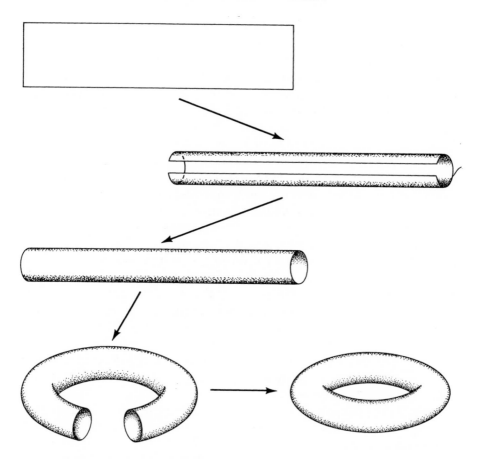

Figure 4.13 A torus is topologically equivalent to a rectangle. Gluing the opposite edges of a rectangle turns it into a cylinder; joining the free pair of edges turns the cylinder into a torus.

construct a three-manifold from any tangled loop of string, no matter how knotted or convoluted the tangle. As in knot theory, that's like confronting two separately snarled masses of fishing line and trying to determine whether or not the two piles of line are tangled in the same way. Unless it's possible to classify such tangles of line in a systematic fashion, there seems to be no hope that three-manifolds can be analyzed either. Solving one problem would solve the other.

Yet even the three-manifold case may be within reach. In 1982, William P. Thurston, then at Princeton University, set forth a program that could lead to the classification of three-dimensional manifolds and a resolution of the original Poincaré conjecture. Thurston's approach uses surgery to reduce complex manifolds to simple cases in much the same way that factoring reduces large composite numbers to products of primes.

Thurston found a geometric pattern that may turn out to encompass all possible three-manifolds and the curved spaces of non-Euclidean geometry. A century ago, topologists believed that only two-manifolds could be described as being flat (having zero curvature), spherical (positively curved), or saddle-shaped (hyperbolic or negatively curved). Three-manifolds look too complicated to fit into such simple categories. Unlike a two-manifold, which can be specified and listed according to its genus, every three-manifold, like a tangled loop of string, seems to have its own distinctive properties and to resist fitting into any larger pattern.

The example of two-manifolds suggests possible ways to tame the unruly three-manifolds. Two-manifolds can be represented as many-sided, or polygonal, figures whose edges are glued together in a particular way. For example, a Flatlander moving on a square may find that when he moves off the top edge of the square, he reappears at the bottom. When the creature moves off the right edge, he reappears at the left. Many video games operate on the same principle to keep a figure on the screen. Such movements can occur if, in some sense, the top of the square is glued to the bottom and the right edge to the left. What does the resulting surface look like? The first gluing creates a cylinder; the second curves the cylinder into a doughnut. The doughnut and the square topologically count as the same manifold—a two-torus (see Figure 4.13).

This gluing trick turns out to be useful in bringing geometry back into the study of three-manifolds, making many manifolds easier to visualize. A three-manifold, for example, can be generated from a rectangular block, such as the space inside a room. Imagine gluing the front wall to the back wall, the left wall to the right wall, and the floor to the ceiling. The result would be a room that has to bend around and join itself in the fourth dimension. All that is needed, however, for the description of the manifold is given by the gluing rules. A human being living in such a house would probably find its geometry quite disconcerting—walking through a door in one wall and ending up on the opposite side of the room, going through the ceiling to come up through the floor, disappearing at one wall and reappearing at another. Because such motion is strikingly similar to a Flatlander's motion on a two-torus, this particular three-manifold is called a three-torus. Abstract gluing shows the close tie between curved three-manifolds and polyhedra such as a cube.

However, at least two major differences hold between the geometry of two-manifolds and the geometry of three-manifolds. First, whereas two-manifolds can have three types of local geometry (flat, spherical, and hyperbolic), three-manifolds can be given five additional types of local geometric structure. In 1980, Thurston proved that most three-dimensional shapes fit into the hyperbolic category and that many exceptions fit into one of the seven other, rarer classes of geometric forms. He conjectured that by looking at small pieces of a manifold, topologists could put each section into one of eight categories, based on the geometric shape of the neighborhood. It's like having eight basic sets of clothing that fit anybody in the world. In a sense, mathematicians impose a geometric structure on a manifold by cutting it into a polyhedron that can then be abstractly glued back together into one of the eight basic geometric types. The trick is to prove that the eight geometric categories suit all conceivable three-manifolds.

Thurston's conjecture has been proved for wide classes of three-manifolds, and it has been tested on many other examples by hand or with the aid of a computer. So far, Thurston's scheme has never failed, and many mathematicians think it's unlikely a counterexample will be found, but no one knows yet for sure.

Thurston's insight has also brought some rigidity back into topology. Normally, topologists study properties that don't change when a shape is deformed. Thurston showed that sometimes it helps to impose a rigid geometric form to pin down topological structures. It makes classification, at least of three-manifolds, easier.

The lack of a proof for the Poincaré conjecture hasn't stopped mathematicians. They work around the conjecture by scrupulously avoiding hypotheses that depend on the truth of the Poincaré conjecture. Such caution is typical of mathematicians. Although most believe that the conjecture is true, a definitive proof is necessary before it can become part of the structure of mathematics. Moreover, as a result of this caution, when a proof finally comes, it won't directly affect much of the work topologists have already done. The proof itself, however, will likely involve new mathematical methods that turn out to be useful elsewhere.

In the meantime, explorations of higher dimensions, strange manifolds, tangled strings, and unruly knots have converged rather suddenly in the last few years to become part of a single story in theoretical physics—one that may cause us to revise our understanding of the role that quantum theory plays in explaining our world.

Tangled Strings

To build a universe like the one we inhabit and observe requires the right sorts of energy and matter interacting in appropriate ways. The so-called standard model of particle physics includes such basic ingredients as quarks, gluons, electrons, and neutrinos. Rooted in the notion of quantum fields, the equations of this theory describe the forces acting between subatomic particles, such as quarks and electrons.

However, there's no place in this framework for gravity. A separate theory and a different set of equations present this fundamental force as a consequence of the curvature of space and time. Einstein's general theory of relativity describes gravity in terms of geometry. Objects travel along the "straightest" possible paths through a dimpled space-time distorted by the presence of mass and energy. Heavier bodies simply create larger dimples and greater curvature.

Theorists have long sought to unite gravity with quantum field theory to produce a "theory of everything." For the past decade or so, the leading candidate for achieving such a unification has been string theory. This theory has just one type of fundamental object. It replaces the many different pointlike particles of the standard model, quantum mechanics, and general relativity with minuscule entities called strings, which can be pictured as closed loops. Although finite in size, these strings of matter and energy are so tiny they look and act like point particles when probed at even the highest energies accessible to particle accelerators.

According to string theory, all matter essentially boils down to infinitesimal strands of energy. Depending on how a particular one-dimensional wisp vibrates or rotates, it manifests itself as an electron, quark, or some other building block of matter. Moreover, these strings can interact only by touching and joining, which automatically makes all forces related to each other.

Solutions to the differential equations of string theory indicate that these strings exist in a 10-dimensional environment. Four dimensions of this peculiar setting—height, width, depth, and time—correspond to the space-time of relativity theory. The remaining six dimensions are somehow crumpled up so tightly that, in effect, they vanish from view. Nonetheless, the geometric properties of these curled-up, compact, six-dimensional spaces are reflected in the physical laws that govern the way matter behaves in everyday four-dimensional space-time.

The trouble is that mathematicians and physicists have discovered thousands of solutions to the equations, each one apparently describing a different six-dimensional structure or space. Such an abundance of solutions leads to an equally large number of possible descriptions of our own four-dimensional universe. Which one matches the real universe?

Mathematicians already have strong evidence of the complexities involved in unraveling these exotic higher-dimensional spaces, even when the dimension is only four. In 1982 the Oxford mathematician Simon Donaldson proved that dimension four, unlike dimensions one, two, and three, may contain manifolds that are topologically but not smoothly equivalent. For some manifolds, no amount of tugging or pushing rids the manifold of all its creases. He also demonstrated that even when four-manifolds can be smoothed out, the process can take many different routes, leading to vastly different results. In essence, Donaldson came up with a series of computable quantities, now called Donaldson invariants, that enabled him to distinguish between four-manifolds that were previous indistinguishable by other means.

Other mathematicians extended Donaldson's work and took it in new directions. They discovered, for example, that four-dimensional Euclidean space can be given innumerable smooth descriptions. In other words, there exist exotic four-manifolds that are topologically but not smoothly equivalent to standard four-dimensional Euclidean space. In all other dimensions, Euclidean spaces have a unique smooth description, which mathematicians have long used and understand well.

Donaldson's results sharpened the view of many mathematicians that four-dimensional geometry is truly more interesting and complicated than they initially had any right to believe. What makes Donaldson's surprising discovery particularly intriguing is that it concerns four-dimensional space—a space that is clearly of physical importance. Physicists suddenly had to face the mind-boggling prospect of having more than one mathematical way to construct what they conceived of as ordinary four-dimensional space-time. The geometric rules governing the four-spaces are the same, but their structures are different.

A crucial hint of how to narrow down the possibilities came out the revolutionary work of Edward Witten of the Institute for Advanced Study and Nathan Seiberg of Rutgers University. Witten and Seiberg in 1994 discovered a way to eliminate certain singularities in a four-dimensional quantum field theory known as supersymmetry—an extension of the standard model of particle physics that attempts to incorporate all the forces of nature except gravity. They did it by introducing to the theory a hypothetical particle called a magnetic monopole. In the theory of electromagnetism, electricity and magnetism are mathematically on the same footing. Thus, pointlike particles having an electric charge (electrons) could, in theory, have magnetic counterparts (magnetic monopoles).

By gradually changing a particular parameter in the equations describing supersymmetry, the theorists could show that this monopole becomes massless right when the equations, in the absence of monopoles, point to infinity as the answer. This approach made it possible to circumvent troubling singularities and obtain reasonable solutions to the equations. It also represented an easier way to do all sorts of other problems involving four-dimensional space, radically simplifying mathematicians' understanding of Donaldson's results and many other aspects of four-dimensional geometry.

For example, computing the Donaldson invariants to characterize four-manifolds was originally a horrendous task. The simpler mathematical apparatus that Witten and Seiberg had developed produced exactly the same results with far less effort. It wasn't long before mathematicians began describing the change as an unprecedented revolution in our way of viewing four-dimensional topology. Indeed, the remarkable equation introduced by Seiberg and Witten not only carried practically all the information contained in Donaldson's invariants but also could be used to derive many new results about four-dimensional manifolds.

However, the picture isn't yet complete. There are hints of an underlying simplicity of breathtaking scope that encompasses a wide swath of geometry and mathematical physics. Isadore Singer of the Massachusetts Institute of Technology remarked: "In the last decade, ideas from quantum field theory have produced startling results in geometries of two, three, and four dimensions. . . . Physicists have obtained new formulas and relationships in mathematical fields far removed from traditional mathematical physics. Geometers are trying to incorporate these novel methods into mathematics."

Mathematicians expect that the mathematical structure underlying geometric quantum field theories will soon be clarified, giving them a powerful instrument for probing low-dimensional geometries. As a result, within a decade they may finally classify all compact, smooth, three- and four-dimensional manifolds. At last, they will then be able to prove the Poincaré conjecture for three-manifolds—that every bounded three-dimensional manifold without holes is just an ordinary three-dimensional sphere.

"We may be as naive about four dimensions today as scientists and philosophers were about two dimensions in 1820 prior to the discovery of hyperbolic geometry," Singer says. "What we now call exotic or fake four-dimensional spaces could give revolutionary new models in cosmology."

One striking feature of exotic four-dimensional spaces is their increasing complexity as their scale increases; they become infinitely complex at infinity. Such a picture may apply to the distribution of matter in the universe. Indeed, as astronomers study the universe on larger and larger scales, some see hints that the distribution of galaxies and interstellar matter doesn't even out. They detect evidence of irregular arrangements of giant structures, punctuated by large gaps, as far as the aided eye can probe.

Perhaps this new four-dimensional structure has no physical meaning, or perhaps it means that the universe we inhabit has a more peculiar structure than anyone ever imagined. In fact, the universe could be as convoluted as a tangled loop of string. Or there may be space-time universes, organized in different ways but coexisting with ours, with which we can't communicate in any way. Startling mathematical results have given physicists much to ponder.

———————— —— —— — ———————— ——

As the mathematician Clifford Taubes noted, "Physics is the study of the world, while mathematics is the study of all possible worlds." Thus, mathematics unveils the infinite possibilities; physics pinpoints the few that structure our universe and our existence.

Euclidean geometry plays a crucial role in the description of physical space, providing a framework for Newtonian mechanics. According to Newton's laws of motion, if there are no external forces, a body moves in a straight line at a constant speed. Any deviation from uniform motion signals the effect of a force, and the amount of the deviation or curvature measures the force's strength. The general theory of relativity links gravity with the geometry of curved space, and new developments in quantum field theory again highlight a profound interaction between physics and geometry.

There are some curves and surfaces, however, that are so convoluted, crinkled, or twisted that the mathematicians who discovered them more than a century ago described them as monstrous or pathological. Remarkably, even these bizarre forms have proved useful in describing important aspects of the physical world, as we shall see in the next chapter.

5

Ants in Labyrinths

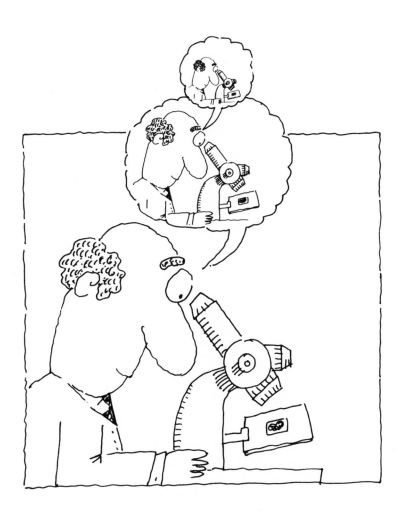

The flickering flames of a campfire highlight the jagged forms of nearby rocks. The smell of frying fish weaves through the still air. Clusters of gnarled pines crouch on the ragged, stony shore. Patches of golden wild flowers, scattered across a meadow, glow in the last rays of a weary sun. At the far end of the lake, a saw-toothed range of towering mountain peaks, crowned with agglomerated snow, tear into the sky. From distant, billowy clouds, a flash of lightning zigzags through the air.

This mountain landscape, like many natural scenes, has a roughness that's hard to capture in the classical geometry of lines and planes, circles and spheres, triangles and cones. Euclidean geometry, created more than 2,000 years ago, best describes a human-made world of buildings and other structures based on straight lines and simple curves. Although smooth curves and regular shapes represent a powerful abstraction of reality, they can't fully describe the form of a cloud, a mountain, or a coastline. In the words of the mathematician Benoit B. Mandelbrot, "Clouds are not spheres, mountains are not cones, coastlines are not circles, and bark is not smooth, nor does lightning travel in a straight line."

The Language of Nature

A close examination of many natural forms reveals that, despite their irregular or tangled appearance, they share a remarkable feature on which we can build a new geometry. Clouds, mountains, and trees wear their irregularity in an unexpectedly orderly fashion. Nature is full of shapes that repeat themselves on different scales within the same object.

A fragment of rock looks like the mountain from which it was fractured. Clouds keep their distinctive wispiness whether viewed distantly from the ground or close up from an airplane window. A tree's twigs often have the same branching pattern seen near its trunk. Elms, for example, have two branches coming out of most forks. In a large tree, this repeated pattern, on ever smaller scales, may go through seven levels, from the trunk to the smallest twigs. Similar branching structures can be seen in the human body's system of veins and arteries and in maps of river systems.

In all these examples, zooming in for a closer view doesn't smooth out the irregularities. Instead, the objects tend to show the same degree of roughness at different levels of magnification. Mandelbrot, the first person to recognize how extraordinarily widespread this type of structure is in nature, used the term "self-similar" to describe such objects and features. In 1975, he also coined the word "fractal" as a convenient label for self-similar shapes. Based on the Latin adjective meaning "broken," it conveyed Man-delbrot's sense that the geometry of nature is not one of straight lines, cir-cles, spheres, and cones, but of a rich blend of structure and irregularity. Fractal objects contain structures nested within one another. Each smaller structure is a miniature, though not necessarily identical, version of the larger form. The mathematics of fractals mirrors this relation between pat-terns seen in the whole and patterns seen in parts of the whole.

Fractals turn out to have some surprising features, especially in contrast to such geometric shapes as spheres, triangles, and lines. In the world of classical geometry, objects have a dimension expressed as a whole number. Spheres, cubes, and other solids are three-dimensional; squares, triangles, and other plane figures are two-dimensional; lines and curves are one-di-mensional; and points are zero-dimensional. Conventional measures of size—volume, area, and length—also reflect this fundamental classification.

Fractal curves can wiggle so much that they fall into the gap between two standard dimensions. Indeed, they can have dimensions anywhere be-tween one and two, depending on how much they meander. If the curve is rather smooth and more closely resembles a line, it has a fractal dimension near 1. A curve that zigzags wildly and comes close to filling the plane has a fractal dimension closer to 2.

Similarly, the fractal dimension of a mountainous landscape can lie somewhere between the second and third dimensions of classical geome-try. A scene with a fractal dimension close to 2 may display a rounded hill with small bumps, whereas one with a fractal dimension near 3 would fea-ture a notably rugged, irregular surface. A higher fractal dimension indi-cates a greater degree of complexity and roughness, but a fractal dimen-sion is never larger than the Euclidean dimension of the space in which the fractal shape is embedded. A hilly scene would never have a dimension greater than 3. In general, fractal geometry bridges the gaps between whole-number dimensions.

Maps of a rugged coastline illustrate another curious property of frac-tals. Finer and finer scales reveal more and more detail and lead to longer and longer coastline lengths. On a world globe, the eastern coast of the United States looks like a fairly smooth line that stretches somewhere be-tween 2,000 and 3,000 miles. The same coast drawn on an atlas page show-ing only the United States looks much more ragged. Adding in the lengths of capes and bays, its extent now measures more like 4,000 or 5,000 miles. Piecing together detailed navigational charts to create a giant coastal map reveals an incredibly complex curve that may be 10,000 or 12,000 miles long. A person walking along the shoreline, staying within a step of the wa-ter's edge, would have to scramble more than 15,000 miles to complete the trip. A determined ant taking the same coastline expedition but staying only an ant step away from the water may go 30,000 miles. Tinier coastline explorers even closer to the shoreline would have to travel even farther.

This curious result suggests that one consequence of self-similarity is that the simple notion of length no longer provides an adequate measure of size. Although it's reasonable to consider the width of a bookcase as a straight line and to assign it a single value, a fractal coastline can't be considered this way. Unlike the curves of Euclidean geometry, which become straight lines when magnified, the fractal crinkles of coastlines, mountains, and clouds do not go away when magnified. If a coastline's length is measured in smaller and smaller steps or with shorter and shorter measuring sticks, its length grows without bound. Because it wiggles so much, the true length of a fractal coast is infinite. Length, normally applied to one-dimensional objects such as curves, doesn't work for objects with a fractional dimension greater than 1.

Fractal geometry doesn't prove that Euclidean geometry is wrong. It merely shows that classical geometry is limited in its ability to represent certain aspects of reality. Classical geometry is still a handy way to describe salt crystals, which are cubic, or planets, which are roughly spherical and travel around the sun in elliptic orbits. Fractal geometry, on the other hand, introduces a set of abstract forms that can be used to represent a wide range of irregular objects. It provides mathematicians and scientists with a new kind of meter stick for measuring and exploring nature.

Taming Mathematical Monsters

In 1904 the Swedish mathematician Helge von Koch (1870–1924) created a mathematically intriguing but disturbing curve. It zigzags so much that a traveler set down anywhere along the curve's path would have no idea in which direction to turn. Like many figures now known to be fractals, Koch's curve can be generated by a step-by-step procedure that takes a simple initial figure and turns it into an increasingly crinkly form.

The Koch curve, which looks like a snowflake, starts innocently enough as the outside edge of a large equilateral triangle (*see Figure 5.1*). The addition of an equilateral triangle one-third the size of the original to the middle of each side and pointing outward turns the figure into a six-pointed star. The star's boundary has 12 segments, and the length of its outer edge is four-thirds that of the original triangle's perimeter. In the next stage, a triangle one-ninth the size of the original is added to the middle of each of the 12 sides of the star. Continuing the process by endlessly adding smaller and smaller triangles to every new side on ever finer scales produces the Koch snowflake. Any portion of the snowflake magnified by a factor of three will look exactly like the original shape.

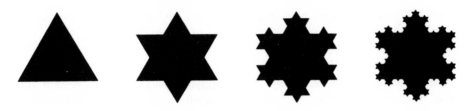

Figure 5.1 The first four stages in constructing a Koch snowflake.

The boundary of the Koch snowflake is continuous but certainly not smooth. So convoluted that it's impossible to admire in all its fine detail, it has an infinite number of zigzags between any two points on the curve. A tangent—the unique straight line that touches the curve at one point—can never be drawn anywhere along its perimeter. Moreover, the length between any two points is infinite, yet the curve bounds a finite area not much bigger than the area of the original equilateral triangle. In fact, despite the figure's wrinkled border, it's possible to compute that the area bounded by the curve is exactly eight-fifths that of the initial triangle.

Such strange mathematical behavior led mathematicians at the end of the nineteenth and beginning of the twentieth centuries to refer to this self-similar curve and several others as mathematical monstrosities. Constructed in the course of attempts to prove or disprove certain intuitive notions about space, dimension, and area, such curves were sometimes described as "pathological." It was mathematics "skating on the edge of reason," as the mathematician Hans Sagan of North Carolina State University remarked.

Nonetheless, such monster curves are remarkably easy to generate. Pointing triangles inward rather than outward—subtracting instead of adding them at each step—produces the antisnowflake curve. This lacy form, too, has an infinitely long outer edge that intersects itself infinitely often, but its area is only two-fifths that of the starting triangle.

The same idea of adding or subtracting pieces that are successively smaller in size works for a square or any other polygon. It also works in three dimensions. Dividing each face of a regular tetrahedron into four equilateral triangles and erecting a smaller tetrahedron on each face's middle triangle, then continuing this step-by-step procedure indefinitely creates a prickly, three-dimensional analog of the Koch snowflake. Its surface area is infinite, but the figure bounds a finite volume. The supply of monsters seems limitless!

No wonder mathematicians were disturbed when they first encountered such bizarre behavior. These pathological curves and surfaces, they believed, were aberrations—skeletons in the closet of otherwise orderly mathematics. To them, such figments of the imagination represented a mathematical pathology having nothing to do with any possible real-world phenomenon and were unlike anything found in nature.

Nevertheless, a few mathematicians took these monster shapes seriously enough to explore their properties in some detail. In 1919, the mathematician Felix Hausdorff (1868–1942) suggested a way to generalize the notion of dimension, which put these disturbing forms into a class of their own. He came up with the idea of fractional dimension, a concept that is now one of several methods used to characterize a fractal.

The idea of fractional, or fractal, dimension extends the concept of dimension normally used for describing ordinary regular objects, such as squares and cubes. The idea is to figure out how many small objects or units of size p are needed to cover a large object of size P. In the case of a line segment, say, 8 meters long, it takes eight 1-meter lengths to cover the whole line. If the measuring unit were 10 centimeters long, it would take 80 such units to cover the 8-meter length. That ratio between the two results, 80 and eight, is $10:1$ or 10^1, which is the ratio of the measuring-stick lengths—1 meter and 10 centimeters. The exponent 1 matches the dimension of a line. Similarly, in the case of area, a square 1 meter by 1 meter fits into an 8-square-meter area eight times, whereas a measuring square 10 centimeters by 10 centimeters fits the same area 800 times. The ratio between the two results, 800 and eight, is $100:1$, or 10^2. The exponent 2 matches the dimension normally associated with area. Volume can be handled in the same way, using measuring cubes of different sizes. In every case, the dimension of the object appears as the exponent of the ratio of the length scale of the measuring units.

The whole process of finding the dimension of an object can be turned into a mathematical operation of taking logarithms of the appropriate ratio. Tripling the width of a square creates a new square that contains nine of the original squares. Its dimension is calculated by taking logarithms of the size ratio, or magnification: $\log 9/\log 3 = \log 3^2/\log 3 = 2 \log 3/\log 3 = 2$. Hence, a square is considered two-dimensional. Doubling the size of a cube produces a new cube that contains eight of the original cubes. Its dimension is $\log 8/\log 2 = \log 2^3/\log 2 = 3 \log 2/\log 2 = 3$. It's no surprise that a cube is three-dimensional.

In general, for any fractal object of size P, constructed of smaller units of size p, the number, N, of units that fits into the object is the size ratio raised to a power, and that exponent, d, is called the Hausdorff dimension. In mathematical terms, this can be written as $N = (P/p)^d$ or $d = \log N/\log (P/p)$. This roundabout way of defining dimension shows that familiar objects, such as the line, square, and cube, are also fractals. The line contains within itself little line segments, the square contains little squares, and the cube little cubes.

Applying the concept of Hausdorff dimension to the Koch curve gives a fractional dimension. Suppose that, at the first stage of its construction, the snowflake curve had been 1 centimeter on each side. If the completed fractal were viewed with a resolution of 1 centimeter, the curve would be seen as a triangle made up of three line segments. Finer wrinkles would be invisible. If the resolution were improved to ⅓ centimeter, then 12 segments, each ⅓ centimeter in length, would become evident. Every time the resolution was improved by a factor of three, the number of visible segments would increase four times. In this case, $N = 4$ and $P/p = 3$. Hence, $3^d = 4$, so log $3d$ = log 4, d log 3 = log 4, and d = log 4/log 3. The Hausdorff dimension of the fractal Koch curve is 1.2618. . . . The strange properties of the snowflake curve stem from the fact that it is not a one-dimensional object. It belongs in the wonderland of fractional dimensions.

Fractals in nature typically lack the regularity evident in, say, a Koch curve, but natural fractals are often self-similar in a statistical sense. With a large enough collection of examples, a magnified portion of one sample will closely resemble some other sample in the collection. The fractal dimension of these shapes can be determined only by taking the average of the fractal dimensions at many different length scales. Although the Koch curve's peninsulas and bays occur with absolute regularity, they bear a striking resemblance to the shape of a rugged coastline. Although the bends and wiggles of a coastline are not in precisely the same locations on all scales, the coastline's general shape looks the same no matter what scale is used for measurement.

Another significant difference between a Koch curve and an actual coastline is that the curve is an idealized mathematical form with structure at infinitely many levels. A physical coastline isn't really an infinitely complex line. However, for many purposes, a fractal is a better, though not necessarily perfect, model of a coastline than is a smoother curve. The similarity between an abstract object like the Koch curve and the typical characteristics of a coastline is enough to define an approximate fractal dimension for actual ocean coastlines. Studying maps on different scales shows that ocean coastlines vary in dimension, but generally range from 1.15 to 1.25—not quite as rough as a Koch curve.

One of the joys of fractal geometry is the opportunity to create new monster curves and other pathological forms. Usually, that means starting with some basic shape or figure (the initiator), then applying a rule (the generator) that step-by-step makes the figure more irregular, tangled, or wrinkled on ever smaller scales in an endlessly looping process.

One strange form, called the Sierpiński gasket, starts as a triangle, like the Koch snowflake, but all the drawing and cutting take place within the figure. Marking the midpoints of the three sides and joining those points creates a new triangle embedded within the original triangle. This construction breaks up the large triangle into four smaller triangular pieces—one inverted central triangle and three corner triangles. Cutting out the central triangle leaves the three corner triangles. The process of finding midpoints

Figure 5.2 This fractal starts as a triangle. Removing successively smaller triangles step by step leads to the remarkable figure known as the Sierpiński gasket, which turns out to have zero area.

and drawing in a triangle is repeated within each of the remaining triangles. The central triangle that sits within each of these corner pieces is cut out, leaving nine small triangles. The pattern seen in the first step is thus duplicated in each of the corner triangles within the figure. The result is a two-dimensional sieve punctured by an infinite number of holes (*see Figure* 5.2).

At each step in the process, the length of a triangle's side is cut in half, and three times as many triangles appear. Therefore, the Sierpiński gasket has a Hausdorff dimension of log 3/log 2 = 1.584. . . . It also happens to have zero area.

A similar procedure can be applied to a square. Dividing each side of a square into three parts to create a three-by-three grid and removing the central square is the first stage in generating an infinitely moth-eaten Sierpiński carpet (*see Figure* 5.3). Its fractal dimension is log 8/log 3 = 1.8928 . . ., showing that this figure is actually more like a bumpy curve than a living-room carpet.

The same operation can take place in three dimensions. Starting with a cube, dividing it into 27 smaller cubes, and removing the central cube as well as the cubes lying at the center of each face of the original cube (seven in all) leads inexorably, step by step, to a novel form called the Menger sponge (*see Figure* 5.4). Its dimension is log 20/log 3 = 2.727. . . . Because its dimension is closer to three than to two, the Menger sponge is more a solid body than a smooth surface.

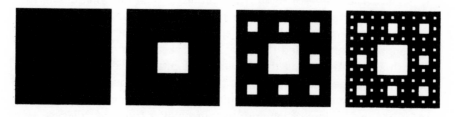

Figure 5.3 A Sierpiński carpet begins as a square. Removing square patches generates the fractal form.

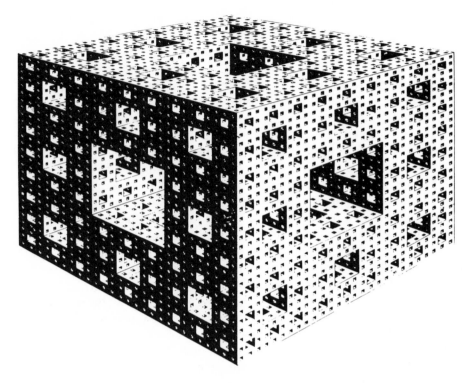

Figure 5.4 A Menger sponge is the three-dimensional counterpart of the Sierpiński carpet.

At the opposite dimensional extreme, a similar removal process can be applied to a one-dimensional line segment. In this case, a section is removed from the middle of a line, then a corresponding section is removed from the middle of the two remaining pieces, and so on, until the line falls apart in a shower of dimensionless fragments. This extremely simple fractal is an example of a Cantor set, named for the mathematician Georg Cantor (1845–1918). All Cantor sets created by breaking up a line have a dimension between zero and one.

Mandelbrot, who initiated much of the research on fractals, has been particularly inventive in generating novel fractals with unique properties. Some of these fractals resemble natural shapes, such as mountains. Others have geometric features that make them useful as idealized models of physical processes, such as the seeping of oil through porous rock or the way in which iron becomes magnetized.

One early application of fractals was to solve the problem of noise during data transmission. Like the irregular crackling sound sometimes heard during a radio broadcast, electrical disturbances interrupt and confuse the flow of data over telephone lines and other transmission channels. By looking at the pattern of errors that occur when computer data are transmitted digitally as a sequence of on/off signals, Mandelbrot noticed that the errors seemed to show up in bursts. Examining these bursts more closely, he found that each burst itself was intermittent. Those shorter bursts and their associated gaps also had a similar structure. The best mathematical model for these characteristic bursts appeared to be a Cantor set.

The Cantor set can be extended to three dimensions to become what Mandelbrot calls a Cantor dust. This fractal dust starts off as a solid block of matter, which is divided into stacks of smaller blocks. Some of the smaller blocks are randomly removed. The remaining blocks are subdivided further, and even more matter is randomly removed. Gradually, this Swiss-cheese structure comes to resemble water droplets scattered in a cloud or even the clusters of stars and galaxies dispersed throughout space. Ultimately, it becomes a dust of totally disconnected points.

Using the Cantor-dust concept, Mandelbrot concocted distributions of stars in a galaxy and galaxies in the universe. Although the model is a forgery and uses no real astronomical data, it closely resembles the hitherto inexplicable distribution of stellar matter seen by astronomers. Astrophysicists have now confirmed that, as the model predicts, mass within the universe is distributed throughout space like a three-dimensional Cantor set, with large regions of space left empty. Cantor-set fractals describe not only the way matter clusters in space but also the way it clusters in time. Cantor sets seem to describe the distribution of cars on a crowded highway, cotton price fluctuations since the nineteenth century, and the rising and falling of the River Nile over more than 2,000 years.

One of the very few scientists who appreciated early on the peculiar geometric complexity of natural phenomena was the physicist Jean Baptiste Perrin (1870–1942), who studied the erratic movements of microscopic particles suspended in liquids and remarked on the self-similar structures of natural objects. As Perrin wrote in 1906:

> Consider, for instance, one of the white flakes that are obtained
> by salting a solution of soap. At a distance, its contour may
> appear sharply defined, but as we draw nearer, its sharpness
> disappears. . . . The use of a magnifying glass . . . leaves us just
> as uncertain, for fresh irregularities appear every time we increase
> the magnification, and we never succeed in getting a sharp,
> smooth impression, as given, for example, by a steel ball.

Generally, however, scientists for a long time ignored the pathological curves and surfaces that mathematicians had unveiled. At the same time, they kept encountering difficulties in answering fundamental questions about physical behavior. Physicists, for example, had trouble explaining aspects of diffusion. When a person in a breezy room opens a bottle of perfume, how long does it take for someone on the other side of the room to smell it? In turbulent air currents, perfume molecules take paths wispier, stringier, and more contorted than the routes they follow in a calm setting. "It's only with Mandelbrot's insight—that all this wispiness involves those pathological nineteenth-century constructions—that physicists were able eventually to start to attack those problems," says Michael F. Shlesinger, a physicist at the Office of Naval Research.

Nowadays, fractals enter into scientists' descriptions of a wide range of phenomena, from the branching of air passages in the lungs and the flight paths of wandering albatrosses to the fracturing of a chunk of metal. They have a potentially important role to play in characterizing weather systems and in providing insights into various physical processes, such as the occurrence of earthquakes and the formation of deposits that shorten battery life. It's even possible to view fractal statistics as a doorway to a unifying theory of medicine, offering a powerful glimpse of what it means to be healthy. "The combination of organization and diversity, observed at all levels of anatomy and control processes in humans, should be described as fractal random processes," argues Bruce J. West of the University of North Texas. He suggests that knowing that a process is, or ought to be, a process governed by fractal statistics may explain the sensitivity an individual has to a particular medication. "It may assist in designing a protocol for the testing of a new drug. It may allow researchers to interpret properly extreme fluctuations in data sets. It may enable us to devise new diagnostics for certain diseases."

———————————— Packing It In ——————————

In his explorations of fractals, Benoit Mandelbrot had a distinct advantage over Perrin, Koch, and other predecessors. He could commandeer computers to calculate and display stunning images of the extraordinary forms. Today, what nineteenth-century mathematicians could barely imagine can be speedily depicted and studied in three-dimensional, full-color splendor.

Fractal geometry and computer graphics are inextricably linked. Computer graphics provides a convenient way of picturing fractal objects, and fractal geometry is a useful tool for drawing computer images. The simple, repeated operations that go into the construction of a fractal are ideally suited to the way a computer functions. The computer patiently and painstakingly performs the same set of operations over and over again to generate a portrait of a particular fractal object.

Of course, the fractal image that appears on a computer screen isn't really, in a mathematical sense, a true fractal. A computer generating a Sierpiński gasket, for example, can display no triangles smaller than the tiny spots of light, or pixels, that make up the screen. The true Sierpiński gasket is filled with even smaller triangles. Computer graphics therefore provides a useful but only approximate rendering of fractal forms.

Fractal geometry overcomes some of the disadvantages that crop up when conventional geometric techniques are used to draw realistic natural objects on the computer screen. One standard conventional approach is to piece the image together from simple Euclidean shapes, including squares, triangles, and circles. Such constructions work well for angular, human-made objects, such as bridges, robots, and spacecraft, but they can't capture the enormous detail and irregularity evident in clouds or natural terrain. Specifying the necessary details would be a monumental computing task.

Fractals offer a simpler solution. The fractal approach to drawing pictures involves generating a basic shape, such as a rectangle. That shape is then recreated many times on smaller and smaller scales within the original figure. Random variations are thrown in to make the image look a little rougher. Such artificial landscapes can be mathematically magnified to reveal more detail, just as a close-up lens probes deeper into a natural scene.

One way to build a mountain using a technique loosely based on fractal geometry is to start with a triangle. The computer first finds the midpoint of each of the triangle's sides. Each midpoint is then displaced along its corresponding edge through a distance determined by a random-number generator. Joining the displaced midpoints generates a new triangle and divides the original into four smaller triangles. Unlike the components of the Sierpínski gasket, these triangles aren't necessarily equilateral or even identical. The same procedure is applied in turn to each of the four new triangles, generating 16 triangles, to each of which the procedure is applied again, and so on. The process continues until the individual triangles are so small that their edges can no longer be distinguished. Although the algorithm for subdividing the triangles is straightforward, the resulting figure is a complex polygonal surface. The addition of color and shadow turns this figure into a reasonable facsimile of a mountain (*see Figure 5.5*).

To generate a fractal landscape, the graphic artist can enter into the computer a set of elevations, specifying the locations of mountain peaks and valleys. The computer then connects the dots to produce a complicated polygon. That figure is further subdivided into a mesh of simpler triangles. Next, the computer subdivides each triangle into smaller triangles, using the same technique as for drawing a single mountain. Given sufficient computing time, the triangular facets can be subdivided to the point where their edges are too small to be distinguished.

This particular recipe for building a mountain landscape is one of the simplest of several possible construction schemes. Mandelbrot's own

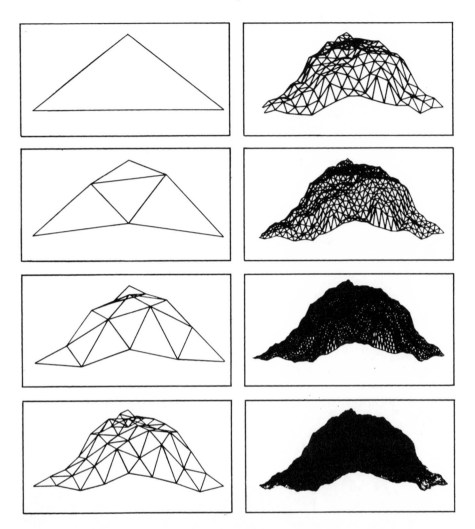

Figure 5.5 One way to draw a fractal mountain is to start with a network of triangles and then break that down into finer nets to generate an irregular surface.

method, which sticks more closely to the true characteristics of a fractal, generates somewhat more realistic scenes—ones that wouldn't be mistaken for mounds of crumpled paper. (The price of this greater realism, however, is longer computation times to generate an image.)

Mandelbrot and his colleagues start with a mathematical construct that closely resembles a random walk—a sequence of up and down steps, determined by the toss of a coin (*see Figure* 5.6). A set of these jagged, vertical cross sections is then assembled to form a markedly rough landscape. A

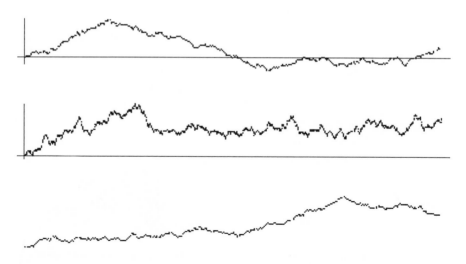

Figure 5.6 Examples of one-dimensional random walks, which can be interpreted as the profiles of mountain ranges.

final mathematical smoothing gives the constructed scene a more natural look. Any points on a random-walk cross section that happen to go below an arbitrarily set zero level are automatically reset to zero. Collections of these points appear as depressions filled with water (*see Figure 5.7 and Color Plate 7*).

Figure 5.7 A fractal mountain like the one shown can be constructed from the contours generated by a random walk.

Clouds can be created by putting together pictures of selected components of white noise, which corresponds to the hiss heard between FM radio stations. White noise, which can be described by fractal geometry, consists of fluctuations spread evenly throughout the radio-frequency spectrum. Selecting a fractal dimension of, say, 3.2 puts together a combination of long-wave and short-wave components that when plotted produce the proper puffiness for a cloud. With the addition of appropriate lighting and coloring, the result is a soft-looking, wispy object.

Fractal coastlines can be generated in much the same way that clouds are created. Both clouds and coastlines have the same basic, computer-generated relief or outline. The key difference lies in the type of lighting applied to the fractal framework. Light reflects directly off the surface of a landscape, but it partially penetrates a cloud and scatters, or softly diffuses. A computer program can mimic these different lighting effects.

Snowflakes can be made by using a fractal branching program to generate a tree structure. The treelike form is then reproduced six times, in the same way that a kaleidoscope creates a faceted display with sixfold symmetry. By changing certain parameters in the branching routine, a wide variety of different snowflake designs can be created (*see Figure* 5.8). However, because a snowflake isn't really a true fractal, the images don't always look quite right.

Creating a fractal, branched structure is fairly easy, but drawing a realistic tree can be quite tricky. Limbs, branches, and twigs all repeat the shape of the tree as a whole on different scales. The routine continues until the smallest twig is a specified fraction of the size of the trunk, at which point the computer program shuts itself off. However, fractal trees drawn in this way always seem more sparse and symmetric than real trees (*see Figure* 5.9, *left*).

One drawback of simple fractal models is that they fail to reflect changes in the branching patterns of real trees as the scale changes. For example, pine needles don't have exactly the same shape as the branches to which they're attached, nor do the branches replicate the shape of the whole tree. Moreover, real trees grow so that their branches and twigs don't overlap.

One promising strategy is to combine the branching character of a fractal structure with just the right dose of knowledge about how plants grow. It turns out that crude assumptions about plant growth lead to remarkably realistic pictures. A formula, with small random perturbations, sets the angle and position of each set of branches. The resulting figure looks much like a natural tree. More sophisticated programs for drawing trees include additional parameters such as the curvature and twist of the branches (*see Figure* 5.9, *middle and left*).

The fractal images just described were originally generated largely by trial and error, perhaps as the result of computer doodlers' stumbling accidentally on mathematical procedures that happen to lead to drawings that look like natural objects. The trial-and-error period, however, is reaching an end as more systematic approaches to creating fractal images are developed.

Figure 5.8 Generating a fractal pattern, then reflecting it six times, results in a form resembling a snowflake.

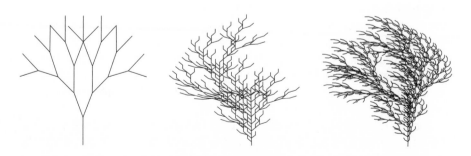

Figure 5.9 Examples of fractal trees, generated using different rules.

In the 1980s, Michael Barnsley, then at the Georgia Institute of Technology, tackled the problem for finding a specific fractal to fit a given natural object. He and his coworkers studied how a scene's geometry can be analyzed to generate an appropriate set of rules that can then be used to recreate the scene. Because fractal mathematics is a compact way to store the characteristics of an object, this approach would compress the content of an image into just a few equations.

The idea is to start with a digitized picture. Such a picture may consist of a thousand-by-thousand grid of pixels. Each pixel is assigned, say, eight bits of data to represent 256 different shades of gray or an equal number of different colors. Thus, the entire picture can be thought of as a string of 8 million 1s and 0s, one digit for each bit. If this string of digits can be encoded in some way to produce a new shorter string of digits, the image is said to have been compressed. Of course, the compressed string should be able to reproduce, pixel by pixel, the original picture.

Barnsley and his colleagues believed that the key to image compression is in the redundancy found in natural forms. For example, one pine needle is more or less like any other pine needle. That means there's really no need to describe every needle one by one. It's sufficient to describe just one. Fractals, in which similar structures are repeated at smaller scales within an object, also capture much of the redundancy found in nature.

"Images contain lots of repetition: You rarely see one piece of grass, one leaf, one edge, one bit of blue sky, one eye, and so on," Barnsley notes. "Any fractal compression system involves a process that looks for these repetitions across scale and position." Then, instead of saving information on every piece of an image, the system merely preserves the prototypical form and how it is used on different scales in different places.

Several important mathematical concepts lie at the heart of Barnsley's scheme. His procedure relies on mathematical operations called affine transformations, originally introduced in 1981 by John Hutchinson for drawing fractals. An affine transformation behaves somewhat like a drafting machine that takes in a drawing—that is, the coordinates of all the points making up the lines in the drawing—then shrinks, enlarges, shifts, rotates, or skews the picture and finally spews out a distorted version of the original. In other words, the transformation "maps" each point in the plane to a new location. The most important mappings for drawing fractals are those that contract the plane by moving points closer together.

Affine transformations can be applied to any type of object, including triangles, leaves, mountains, ferns, chimneys, clouds, and even the background scene against which an object may appear. In the case of a leaf, the idea is to find smaller, distorted copies of the leaf that when fitted together and piled one on top of another so that they partially overlap form a collage, which approximately adds up to the original, full leaf. Each distorted, shrunken copy is defined by a particular affine transformation, or contractive map, of the whole leaf. If it takes four miniature copies of the leaf to approximate the whole leaf, there will be four such transformations (*see* Figure 5.10, *top*).

Now, the original image, or "target," whether leaf or cloud, can be thrown away, leaving only the corresponding collection of affine transformations. These can be used to recreate the original image, essentially by molding a piece of space. That's done by starting with a point somewhere on a

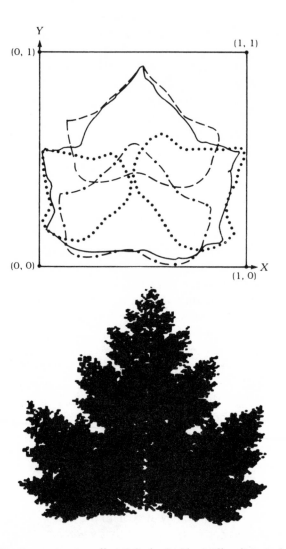

Figure 5.10 *Top*: Approximate collage of a leaf with smaller distorted copies of itself. *Bottom*: An image of a leaf generated by randomly applying a set of affine transformations.

computer screen and applying one of the available transformations to shift the point to a new spot. That spot is marked. Again, randomly applying one of the transformations shifts the point to another location. The new spot is colored in, and the process is repeated again and again. Amazingly, although the designated point appears to hop about aimlessly, first pulled one way then another, a pattern gradually emerges. The point's colored tracks add up to an image called an attractor—a concept discussed in more detail in Chapter 6. In the case of the four leaf transformations, the attractor is an object that looks very much like the original leaf (*see Figure* 5.10, *bottom*).

How this part of the process works can best be understood using a simple example on a sheet of squared graph paper. Start with a triangle with vertices labeled 1, 2, and 3. Mark a starting point anywhere within the triangle. Three rules (transformations) specify where subsequent points fall. The roll of a die or some kind of random-number generator specifies which rule to use. If the number 1 comes up, the new point falls halfway between the previously specified point and the triangle vertex labeled 1. If the number 2 comes up, the new point falls halfway between the previous point and vertex 2. If 3 comes up, the new point falls halfway between the previous point and vertex 3.

At first, the points seem randomly distributed, but after 100,000 or so turns, a definite pattern begins to emerge. In this particular example, the figure turns out to be not a randomly distributed dust of points but the highly regular form of a Sierpiński gasket (*see Figure* 5.11). Different sets of rules produce different fractals. With enough rules, a computer can generate not only fractal curiosities but also convincing images of natural objects, such as leaves, ferns, clouds, and forests.

It turns out that any particular collection of affine transformations, when iterated randomly, produces a unique fractal figure. The trick is to find the right group of transformations to use for generating a particular image. That's done using the collage process, for example, a leaf covered by little copies of itself. Furthermore, the probability of using a certain transformation need not be the same as the probability of applying any other transformation in the set. And because some grid squares are likely to be visited more often than others, keeping track of the relative number of visits to each square provides a way to specify color brightness and intensity or to define a gray scale. In this way, a lot of information is packed into a few formulas, and you can compress images by encoding them as a collection of rules. Randomly iterating the rules then recreates the image.

Using his technique, Barnsley and his group have been able to create remarkable three-dimensional renderings of natural objects such as ferns. Their graceful model of a black spleenwort fern (*see Figure* 5.12) is the product of the application of a collage of four affine transformations—each a combination of a translation, a rotation, and a contraction. When applied to a given point (x, y), a particular transformation generates a new point:

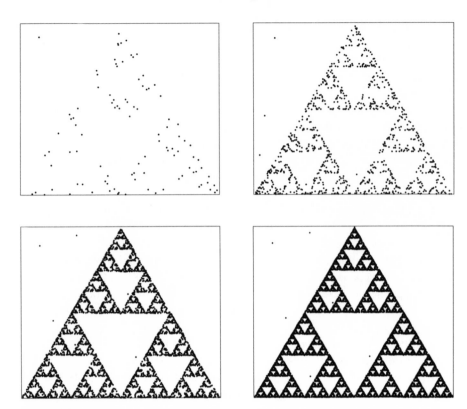

Figure 5.11 Repeated applications of a set of affine transformations generates the Sierpiński triangle, which becomes more and more evident as the number of points increase.

$[rx(\cos A) - sy(\sin B) + h, rx(\sin A) + sy(\cos B) + k]$. The parameters h, k, A, B, r, s have the values shown in the following table:

	Translations		Rotations		Scalings		
Map	h	k	A	B	r	s	Probabilities
1	0.0	0.0	0	0	0.0	0.16	.005
2	0.0	1.6	−2.5	−2.5	0.85	0.85	.8
3	0.0	1.6	49	49	0.3	0.34	.0975
4	0.0	0.44	120	−50	0.3	0.37	.0975

Figure 5.12 A set of simple affine transformations instructs a computer to generate this image of a black spleenwort fern.

Early on, Barnsley and his team were able to come up with reasonable fractal reproductions of various photographs. In one instance, he used 57 affine transformations, or maps, as they are often called, and four colors—a total of two thousand bits of information—to model three chimneys set in a landscape against a cloudy sky. "The idea is that we can fly into this picture," Barnsley said. "You can pan across the image, you can zoom into it, and you can make predictions about what's hidden in the picture."

As you blow up the picture to show more and more detail, parts of the image degenerate into nonsense, but some features, such as the chimneys, the smoke, and the horizon, remain reasonably realistic, even when the image compression ratio is more than 10,000 to 1. The degree of compression attained depends on how much of the regenerated picture makes sense.

The images created using this scheme aren't actually self-similar. Instead, they're referred to as self-affine because an object such as one of Barnsley's fractal ferns shows slightly different features on different scales. Magnifying the image reveals subtle differences in form and color, and the magnification can be carried on indefinitely, as for any fractal.

One of the most striking applications of Barnsley's scheme, which is known as an iterated function system, for achieving image compression is on display in the multimedia encyclopedia *Encarta*, published by the Microsoft Corporation. Stored on one CD-ROM, the encyclopedia includes 7,000 color photographs, which can be viewed interactively on a computer screen. The images range from buildings and musical instruments to people's faces and baseball bats. Normally, each image would take up a huge amount of storage space, but fractal compression makes the task manageable.

At present, fractal image compression works better in some instances than others. Barnsley says, "Current fractal image compression is most effective on real-world images in multimedia and online situations, where the speed of image delivery and decompression are important and where the time taken to compress images is not an issue." In other words, one can call up an image quickly, but creating the compressed form in the first place can be rather time-consuming. Eventually, it may even be possible to convey a movie from one computer to another simply by sending a chain of formulas down a telephone line.

Fractal Excursions

When the delicate fragrance of a perfume weaves through the air, individual perfume molecules, jostled about haphazardly in the hurly-burly of their collisions, follow a remarkably jagged path. Air currents further tangle these tortuous paths into monstrous trails. This type of irregular motion can also be seen in the unceasing, restless dance of tiny, barely visible particles suspended in a liquid. Such jitteriness is known as Brownian motion.

In principle, a researcher can trace a Brownian particle's irregular trajectory by periodically plotting the particle's precise location. The result is a string of points strewn across a sheet of graph paper (*see Figure* 5.13). Using straight lines to connect the picture's dots, however, fails to capture the motion's true intricacy. Between every pair of plotted points lies another jagged path.

Perrin and his students conducted just this experiment in the early part of the twentieth century, using a microscope to track the movement of spherical Brownian particles. Their plots showed a highly irregular track, yet they gave "only a very meager idea of the discontinuity of the actual

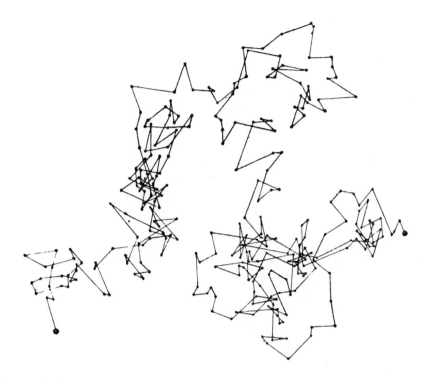

Figure 5.13 A particle exhibiting Brownian motion zigzags randomly along a path that closely resembles a fractal.

trajectory," Perrin noted. If the researchers had increased the resolving power of their microscope to detect smaller effects, they would have found that parts of a path that initially appeared straight would themselves have had a ragged structure.

This type of motion is akin to a random walk—a sequence of steps whose size and direction, as we saw earlier, are determined by chance. In a simple, one-dimensional version of a random walk, a person flips a coin and takes one step forward if the result is heads and one step backward if the result is tails. Across a broader stage, a random walk may look like the steps of a drunken sailor.

When a Brownian trajectory confined to the plane is examined increasingly closely, its length, like that of a coastline, grows without bound. In fact, the trail itself ends up filling practically the whole plane surface on which the motion takes place. Although the trajectory itself is a one-dimensional curve, the path's tendency to fill the plane marks it as a fractal of dimension two.

Fractal concepts can be used to describe not only an astonishing array of fragmented or branching natural structures but also the dynamic properties of these structures—from the movement of Brownian particles to the drip of scalding water through coffee grounds. In a way, fractals represent a new kind of meter stick that scientists can use to measure natural phenomena. For example, they can use fractals to study the way materials are put together, the way they shatter, the way they branch, and the way they conduct heat or electricity.

A word of caution is warranted, however. Mathematical fractals have properties that aren't actually found in natural objects. No real structure can be magnified repeatedly an infinite number of times and still look roughly the same. One reason is the finite size of atoms and molecules. Another is that real objects may, at some magnification, abruptly shift from one type of structural pattern to another. Nevertheless, fractal models provide a useful approximation of reality, at least over a finite range of scales.

The microscopically jagged surface of a chunk of fractured metal is one example of a material property that lends itself to fractal analysis. In one pioneering effort, Mandelbrot worked with some metallurgists to come up with a method for specifying the roughness of a given surface. They found that a large number of broken metal surfaces, though not all, have a roughness that can be represented by a fractal dimension. Experiments showed that the measured fractal dimension takes on the same value for different specimens of identically treated samples of the same metal. The investigators also discovered that different heat treatments not only affect the toughness of a metal but also change its fractal dimension. They concluded that a metal surface's fractal dimension may itself be a useful measurement of a metal's toughness or strength, providing metallurgists with a new tool for characterizing metals.

Fractal ideas have also encouraged scientists to look anew at old, seemingly inexplicable experimental results once destined for the wastebasket and to reexamine problems that previously looked so complicated that they were ignored. Physicists and other researchers now realize that many formerly puzzling results actually reflect the dimensions of fractal geometric objects. With this insight, some apparently complex problems become relatively simple.

In analyzing experimental data, scientists usually look for simple relationships between variables, such as the relationship between the intensity of sound waves scattered from a metal surface and the waves' frequency. If a theory predicts that doubling the frequency will quadruple the intensity, the intensity should be proportional to the frequency squared. However, in many experiments the exponents that express the proportionality turn out to be numbers like 2.79 instead of integers. Scientists who were taught to think of integers as the natural way to represent physical processes are only now beginning to see that noninteger exponents are just as likely, if not more likely, to turn up in nature.

An example from the world of electrochemistry shows the interaction between experimental observation and a mathematical model. For decades, scientists had noticed that the interface between a metal electrode and the electrically conducting liquid, or electrolyte, in which the electrode is immersed has unexpected electrical properties. Simple electrical theory predicts that the interface should behave like a capacitor, a device that stores then abruptly releases electric charge. When an alternating current passes between electrode and electrolyte, it ought to meet a resistance that is simply related to the frequency. However, experimental studies show that this resistance is actually inversely proportional to the frequency raised to a fractional power between zero and one.

Further study shows that the resistance appears to depend on the roughness of the electrode surface. As the surface is made smoother, the exponent approaches one. Under magnification, however, even well-polished electrode surfaces still show long grooves with jagged cross sections and edges. The grooves themselves have finer scratches, suggesting a degree of self-similarity. The theorist's job is to come up with a suitable mathematical model that captures an electrode's observed characteristics and its consequent behavior. That model must be simple and regular enough to be mathematically solvable but not so far removed from an electrode's observed features that any mathematical results from applying the model would be suspect. The idea is not to paint a realistic portrait but to capture the spirit of the phenomenon with a caricature. In that way, researchers may catch a glimpse of what is actually happening at the interface.

Given the nature of the observed electrode grooves, a good place to start is with a fractal. But which fractal? There are many to choose from. Some are easier to handle mathematically than others. One good fractal candidate, related to the Cantor set described earlier, is called the Cantor bar and has a dimension less than one.

The Cantor bar begins as a thick, solid line segment. Taking a chunk out of the center of the bar breaks it into two pieces. The two resulting fragments are in turn broken according to the same rule, and the process is repeated ad infinitum (see Figure 5.14, left). If the length of each broken piece is $1/a$ times the size of the original piece, where a is greater than 2, the object has a fractal dimension of $\log 2/\log a$, which must be less than one.

When the increasingly fragmented bars are welded together—starting with the full original bar on the bottom, then adding the next two pieces, then topping those with the next four broken pieces, and so on—the result is a kind of symmetric urban landscape with stepped towers stretching higher and higher as they get thinner and thinner (see Figure 5.14, right).

Figure 5.14 *Left*: The first five members of a Cantor bar. *Right*: A Cantor-bar model of a rough surface.

Although the surface of an electrode isn't nearly this regular, the mathematical model captures some of the features of a rough grooved surface, especially the idea of grooves within grooves within grooves.

Giving electrical properties to this geometric pattern allows the calculation of various quantities that can also be measured experimentally. Currents, for example, can pass directly across the interface or out along the grooves. That leads to a simple equation predicting that the interface resistance is inversely proportional to the frequency raised to a certain power, and the exponent depends on the geometry of the surface. Now, it's up to experimentalists to see how well this fractal theory matches what's seen in the laboratory. The theorist would then be able to refine the model further and to suggest which measurements would more likely reveal the processes occurring at the electrode-electrolyte interface.

Fractal models also play a useful role in describing percolation processes. The word "percolation" conjures up an image of brewing coffee or the trickle of liquid through a gravel bed, but it also applies to an important class of structures known as percolation clusters. Such clusters have properties similar to those of a floor covered with a random mixture of copper and vinyl tiles. Current will flow from one side of the floor to the other if there is a continuous copper path, no matter how roundabout. If most of the tiles are vinyl, current isn't likely to pass all the way through. Increasing the proportion of copper tiles raises the probability that patches of copper tiles will be linked to form a continuous path. A percolating cluster represents the point at which a conducting path is first created. Percolation clusters are particularly useful for modeling processes such as the seeping of oil through porous rock and the spread of plant species in a meadow.

This construct seems to work as a mathematical metaphor for many materials in which different substances having antagonistic properties are intermingled. Such materials may contain random mixtures of electrical conductors and insulators, magnetic and nonmagnetic components, or elastic and brittle parts. The critical concentration of the two materials in the cluster, not the simple average of the properties of the components, determines which property predominates and how the mixture behaves. This in turn decides physical characteristics, such as the structure of some polymers, the conductivity of alloys, the efficiency of telephone networks, the propagation of forest fires and infectious diseases, and the spread of plant species in an ecosystem. Percolation models also provide useful pictures of abrupt changes in phase, such as the lining up of atomic spins to create a magnet or the sudden temperature-dependent onset of superconductivity in a thin metal film.

Not only are the results of calculations done on an actual random percolation cluster hard to understand, but the computations use up a great deal of computer time and are only approximate because of the complexity of the process. Researchers therefore look for fractal patterns that are systematic enough to ease mathematical calculations but random enough to be realistic.

Fractals also come up in percolation problems in another way. In a problem first proposed in the sixteenth century, an orchard is laid out in the shape of a checkerboard, with each space occupied by a tree. The spaces are small enough that neighboring trees touch each other. Tragically, a plague invades the orchard and spreads from tree to tree, destroying the entire fruit crop. The question at the time of replanting is how many trees should be excluded from the grid to prevent a disease from spreading from one end of the orchard to the other, while still maximizing the fruit yield.

Over the years, various attempts to answer the question have led to the best estimate that about 41 percent of the checkerboard squares should remain unoccupied. The critical concentration for percolation (when the disease can make its way from tree to tree across the orchard) is 59.2 percent. At this threshold value, the pattern of trees occupying the orchard is self-similar, from the overall pattern spanning the orchard down to the individual tree. Above the percolation threshold, the tree cluster is self-similar only on certain scales. On other scales, the pattern is merely some Euclidean object.

Another intriguing and scientifically rewarding pursuit is to look at fractals upon fractals. Diffusion, as exemplified by the peregrinations of perfume molecules, is a fractal process. When diffusion occurs along a fractal surface, the process is something like the wanderings of an ant in a labyrinth (see Figure 5.15). An ant constrained to roam along a straight line always returns to its starting point eventually. On a two-dimensional plane surface, the ant gets lost. However, on fractal paths with dimensions between one and two, what happens to the ant isn't completely known. Depending on the nature of its fractal labyrinth, the

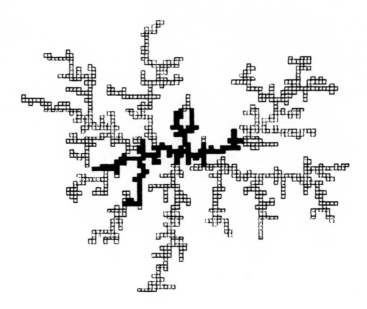

Figure 5.15 Visited sites are shown in black in this 2,500-step random walk on a 1,000-site fractal substrate.

ant may keep running into dead ends forever, or it may return to its starting point very infrequently.

Sticky Stuff

One of the wonders of nature is how simple water molecules can settle into the symmetrical structures that have the limitless variety and intricacy of snowflakes. Very little is known about how these lacy shapes are created. Why does a snowflake branch? How does one branch tell another which way it's going so that the whole flake stays more or less symmetrical?

To try to account for how a snowflake or other growing body flowers into its final form, scientists have developed a variety of simple mathematical models that suggest ways in which growth can occur. By studying these models and comparing them with experimental observations, they can begin to guess what forces and conditions underlie various types of growth.

One of the simpler growth models, out of which a fractal form springs, starts as a tiny cluster, or "seed." Every time a particle wandering by happens to bump into the cluster, it sticks and stays put. Once it has arrived, the particle never jumps to another site. This type of process is called aggregation. If particles diffuse to the cluster by means of random walks—ordinary Brownian motion in three dimensions—the process is diffusion-limited aggregation.

This growth process can be simulated in two dimensions, step-by-step on a computer. At the center of a grid sits a small, connected set of dots representing the initial aggregate. Each dot lies in a separate square of the grid. Far away from the central cluster of dots, a single particle starts its random walk. When the walker eventually arrives at an unoccupied space neighboring a square already filled, it stops and stays put, and the cluster grows by one unit. The process is continued for perhaps 50,000 or 100,000 wandering particles.

The resulting pattern is a typical fractal object, which looks like a bare tree seen from above, with branches shooting off in all directions (*see Figure 5.16, left, and Color Plate 8*). This particular fractal object also contains tremendous empty regions, often in the form of long narrow channels, or fjords, that penetrate far into its interior. What accounts for this distinctive structure is the fact that few late arrivals can get deep into a fjord. Random walkers that happen to reach a fjord's outlet and begin their journey down the channel are much more likely to bump into a wall than they are to make it all the way to the end.

In an effect called growth instability, any cluster that is slightly distorted initially will become even more distorted. Once the initial cluster begins to develop bumps and hollows, the bumps end up growing much faster than the hollows. Because particles are more likely to stick near peaks, the

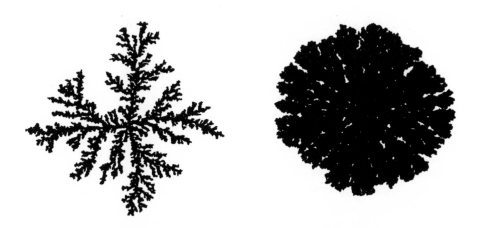

Figure 5.16 *Left:* A 100,000-particle cluster grown on a square lattice using the diffusion-limited aggregation model. *Right:* A 180,000-particle cluster formed in a two-dimensional simulation of ballistic aggregration.

bumps grow even steeper, and the fjords become less and less likely to fill. Eventually, continual growth and splitting of the protruding ends give rise to a highly branched globular fractal.

What happens if a particle sometimes bounces off instead of sticking? Computer simulations show that this leads to a thickening of the branches and an increase in fractal dimension, but the figure remains a fractal.

Although diffusion-limited aggregation is easy to describe and simulate, the process is not yet well understood at a deeper level. No one is sure why the process gives rise to fractals rather than shapeless blobs that have no symmetry at all. Why are loops—or open lakes surrounded by matter—so rarely formed? How does the fractal dimension depend on the dimension of the space in which the process occurs? Ordinary mathematical tools seem inadequate for answering these and related questions.

The diffusion-limited aggregation model can be used, nevertheless, to explain certain natural growth processes, such as the deposition of metals during an electrochemical reaction (*see* Figure 5.17). Not only do many natural forms have fractal characteristics, but fractals also serve as models of new materials that don't exist naturally. For example, diffusion-limited aggregation leads to materials with an especially large surface area. Experimentalists have created one such material by letting microscopic gold balls diffuse toward a small cluster, which grows into a three-dimensional aggregate that's so branched it's practically all surface.

What about snowflakes? Diffusion-limited aggregation theory has made a good start on reproducing the branched, hexagonal laciness of snowflakes. A sixfold pattern is easily built in when one lets the particles diffuse on a triangular lattice rather than a square grid. A controlled touch of randomness, or "noise," decides which of many equally probable sites should grow at each step. The resulting figures look like snowflakes, although they lack the amazing similarities shown among the branches of different arms in a real snowflake.

Another way to create a growth model is to begin a computer simulation with many particles distributed throughout space and to allow the particles to move about randomly until they meet and stick together. In this case, the clusters can also move, which leads to a different type of pattern with a fractal dimension of about 1.4.

In ballistic growth, a given particle moves not on a random walk but along a randomly aimed straight line. If it strikes a predecessor, it sticks where it hits. The process generates patterns in the plane that after a few thousand trials look somewhat like porous, deckle-edged spots roughly circular in shape (*see* Figure 5.16, *right*). Simulations show that if growth goes on for long enough—through a quarter of a million or so trials—the laciness dwindles, and the covered area grows with the square of the radius,

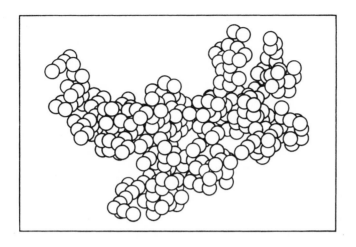

Figure 5.17 *Top:* Diffusion-limited aggregation, simulated in a three-dimensional space by a computer, gives rise to a fractal with a dimension 2.4. *Bottom:* The pattern is similar to that of a copper cluster.

while the edges become better defined. The fractal dimension approaches the Euclidean value of 2.

However, in most fractal growth models, mathematically precise results remain elusive because so many of their properties aren't known or understood. Even the concept of fractal dimension isn't enough to distinguish among the diverse objects that appear to qualify as fractals. To overcome this deficiency, various other types of dimensions and constants have been introduced for special cases. Mathematicians and scientists have also been searching beyond the fractal dimension for other properties that may be universal.

A startling result of current research on fractal-modeled physical processes is that very different phenomena give rise to fractal dimensions that are very close in value. Such events range from the distribution of galaxies in the universe to the distribution of whirlpools in turbulently flowing fluids. It's too early yet to tell whether these phenomena are a collection of separate problems with a different explanation for each case or whether they depend on some underlying principle that would explain many of them simultaneously.

Ever since the earliest applications of fractal geometry, description has outpaced explanation. Theories that depend on fractal properties seem to work, but no one really knows why. Further progress in the field depends on establishing a more substantial theoretical base in which geometric form can be deduced from the mechanisms that produce it. Without a strong theoretical underpinning, much of the work on fractals seems superficial and perhaps pointless. It's easy to perform computer simulations on all kinds of models and compare the results with each other and with experimental results. Without organizing principles, however, the field drifts into a zoology of interesting specimens and facile classifications.

Meanwhile, the use of fractals as a descriptive tool is diffusing into more and more scientific fields, from cosmology to ecology. To some extent, fractals are in the eye of the beholder. Just as a river system may be identified by naming it as a whole or by naming each of its numerous tributaries, researchers can choose different ways to describe the objects they see in nature. They can label the branches individually or recognize that a branch looks roughly like the whole object and call it a fractal. Time will tell whether it's useful to see objects in terms of fractals. A meter stick is a very humble object until it gets into the hands of an Einstein. Albert Einstein's work with meter sticks and clocks led to profound changes in our concepts of space and time. Can a fractal meter stick be used to construct a deeper theory of why fractal objects occur, why certain fractal quantities are universal, and why fractals are ubiquitous in nature? This new meter stick has yet to find its Einstein.

——————— —— —— — —————— ——

Fractals have invaded the popular imagination. Calendars, computer screens, and books feature vivid, phantasmagorical images of weirdly branched, wildly swirling structures. The cartoonist Sidney Harris depicted a refurbished living room decorated with squiggles. "We did the whole room over in fractals," the hostess explains. The composer-pianist Zach

Davids turned the subtly varying fractal intervals between successive heart-beats into musical sequences to create an album of unexpectedly soothing music without apparent rhythm, meter, or harmony. A character in Tom Stoppard's 1993 play *Arcadia* asks, "If there is an equation for a curve like a bell, there must be an equation for one like a bluebell, and if a bluebell, why not a rose?"

Fractals lie at the heart of current efforts to understand complex natural phenomena. Unraveling their intricacies could reveal the basic design principles at work in our world. Until recently there was no word to describe fractals. Today, we are beginning to see such features everywhere. Tomorrow, we may look at the entire universe through a fractal lens.

6

The Dragons
of Chaos

With scarcely any warning, the human brain can fail. When that happens, a victim can spend seconds staring blankly into space or, in extreme cases, shudder violently, lose consciousness, and fall stiffly to the ground. Such seizures, or epilepsies, are symptoms of overactivity among the brain's nerve cells. Somehow, an electrical disturbance originating in a few neurons temporarily takes over the brain, and no other messages get through.

Mathematical techniques now being developed may provide some insight into the causes of seizures and perhaps suggest a basis for controlling them. The same mathematical methods can also be applied to chemical reactions, electronic circuits, lasers, computer networks, and other systems. The mathematical models indicate that, just as an electrical imbalance can generate a disruptive signal in the brain, slight changes in key components may also trigger unpredictable, erratic activity in other systems.

Steps in Time

The thread that ties together irregular behavior observed in chemical, electrical, mechanical, and physiological systems is an emerging field of mathematics known as nonlinear dynamics, which describes the way in which systems change over time. This mathematical approach suggests that systems governed by strict physical laws can nonetheless shift into a highly irregular form of behavior now called chaos.

First applied in 1975 by James Yorke, a mathematician at the University of Maryland, chaos refers to the apparently unpredictable behavior of a deterministic system governed by mathematically expressed rules. Thus, like many terms appropriated by scientists and mathematicians to encapsulate new theoretical concepts, the word "chaos" takes on a specific meaning in a scientific and mathematical context. The technical definition of chaos, somewhat tenuously related to its everyday meaning, carries with it an image of order in the midst of disorder. That definition contrasts with normal usage, wherein we mean by chaos either a state of utter confusion or a state in which chance is supreme.

Chaos shows up in many physical systems, especially those that involve turbulence—from the way in which water in a mountain stream swirls and splashes down a rocky channel to the manner in which smoke rising in still air starts as a well-defined plume, then gradually breaks up into a disorderly tangle of smoky threads.

Some motions, such as the trajectory of a baseball, are inherently predictable. A good fielder intuitively knows where the ball is likely to fall and automatically uses that information to catch it. On the other hand, the path of a flying balloon expelling air is erratic and unpredictable. The balloon lurches and turns at times and places impossible to predict. Nevertheless, both balloon and baseball obey Newton's laws of motion. Somehow, predictability and unpredictability together reside in the same set of equations. That contradicts the venerable notion that, given the initial position and velocity of each particle in a system and applying Newtonian mechanics, the system's history can be precisely predicted forever. The idea works reasonably well (though not perfectly) for planets in the solar system but not so well to the way ocean waves crash into a beach or the erratic fluctuations in current encountered in a noisy electronic circuit.

The key notion is that the behavior of a chaotic system tends to change drastically in response to a slight change in initial conditions. A double pendulum, for example, can display wildly erratic gyrations (*see Figure 6.1*). Two identical double pendulums starting at slightly different positions very quickly go out of sync and embark on highly individual movement patterns.

Understanding chaotic behavior usually means tracking the system's activity from one moment to the next. Mathematically, it means closely examining the solutions of differential equations that describe and model the phenomena scientists encounter in nature. In mathematical models of neural systems, the variables could represent, for example, the electric potential (or voltage) across cell membranes, membrane currents, and the concentrations of certain chemical substances. Some parameters, such as temperature or a particular drug concentration, assume a constant value for a given case but may vary from case to case.

A differential equation links the rate of change over time of one of the variables with the variable's present value and the present value of other variables. For example, the swinging of a pendulum is governed by an

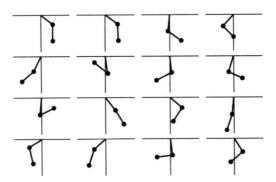

Figure 6.1 A double pendulum, in which one rod is pivoted from the bottom of the other, can show a wide range of unpredictable, capricious movements.

equation in which the only variable is the angle of the pendulum measured from a vertical position. Solving the equation means determining how that angle changes as the pendulum oscillates back and forth.

Descriptions of complicated systems may require a large set of such equations. In a sense, a differential equation is like a machine that takes in values for all the variables and then cranks out new numbers corresponding to the values of those variables at some later time. Often the relationship expressed in the equations is nonlinear (as in the case of the oscillating pendulum), that is, input and output are not proportional.

Mathematicians have learned that under the right conditions, even simple nonlinear differential equations can yield sequences of numbers that appear to follow no pattern. Although the equations express direct cause and effect relationships, the numerical results predict that the modeled systems can show irregular motion or randomlike, chaotic behavior. In fact, this class of solutions displays a sensitive dependence on initial conditions, so that a slightly different starting point produces a radically different result. In principle, the future is completely determined by the past. In practice, however, small uncertainties are amplified, so that even though the behavior is predictable in the short term, it is unpredictable in the long term.

Such irregularity can have startling consequences. If weather systems can be described by mathematical equations that shift into chaotic behavior, a change as slight as a butterfly flapping its wings near a weather station would make long-term weather prediction impossible. In general, the existence of chaotic regimes implies new fundamental limits on predictability, if not a theoretical excuse for the unreliability of today's local weather forecasts. At the same time, the determinism inherent in chaos suggests that many random phenomena are more predictable than had been thought.

The mathematical side of chaos is seen most easily in phase space, where, as discussed in Chapter 4, each dimension represents one of the variables in the differential equations used to model a particular system. Researchers are interested in what happens to phase-space trajectories for different equations and under various circumstances.

Consider a mass oscillating on the end of a spring. If there were no friction, the spring's motion would continue forever, and its phase-space portrait would be a single loop along which the motion endlessly circulates. If instead it gradually loses its energy and stops, the motion, as seen in phase space, settles down and is sucked into one point and stays there. Its phase-space portrait, with position plotted against velocity, looks like a spiral that gets tighter and tighter, ending up at a fixed point. That fixed point is another example of an attractor.

A pendulum clock, in which energy lost to friction is replaced by a mainspring or a system of falling weights, cycles periodically through such a sequence of states. The same principle applies to metronomes and, more or less, to the human heart—they repeat the same motion over and over again. In phase space, such a motion corresponds to a cycle or periodic orbit. Such attractors are called limit cycles. No matter how the motion is started and despite initial fluctuations, the cycle approaches the same long-term limit.

A particular system may have several attractors. Different initial conditions drive the system to a different attractor. The set of points going to the same attractor is called a basin of attraction. Attractors themselves can have a variety of geometric shapes, from something as simple as a torus to complicated, higher-dimensional forms. Nevertheless, motion on all the attractors discussed to this point is completely predictable, so it is non-chaotic.

Under certain conditions, however, nonlinear differential equations generate trajectories in phase space that form peculiar shapes, having none of the regularity associated with the previous examples of attractors. Such objects are sometimes called chaotic, or strange, attractors. Trajectories skip about on the surfaces completely unpredictably. A slight change in the initial conditions greatly changes the particular orbit generated.

In 1963 the meteorologist Edward Lorenz of the Massachusetts Institute of Technology discovered a strange attractor resulting from a simplified set of differential equations describing air flows in the atmosphere. That attractor is now known as the Lorenz butterfly attractor because its curious shape, rendered in two dimensions, resembles the flapping wings of a butterfly (see Figure 6.2). As with all other strange attractors, the behavior of a trajectory on its surface is totally erratic and unpredictable, but the trajectory never leaves the attractor's surface. In weather terms, Lorenz's attractor places limits on what the weather can do. For example, it's not possible to have a blizzard in the Sahara Desert or a 200-degree heat wave in New York City. Within certain limits and given enough time, however, the weather can do practically anything. No matter how much data on temperature, barometric pressure, and wind direction are collected, the weather's true behavior over time veers dramatically from predictability.

Since Lorenz's discovery, mathematicians and other researchers have discovered additional strange attractors, which crop up in various nonlinear differential equations. Chaotic attractors have also been observed in several experiments related to fluid flows and chemical reactions. Among them are the convection pattern of fluid heated in a small box, oscillating concentration levels in a stirred chemical reaction, the beating of chicken heart cells, and a large number of electrical and mechanical oscillators. In addition, computer models of events ranging from disease epidemics to electrical activity of a nerve cell to stellar oscillations have expressed this simple type of randomness. There are even experiments that have searched for chaos in areas as disparate as brain waves and price fluctuations.

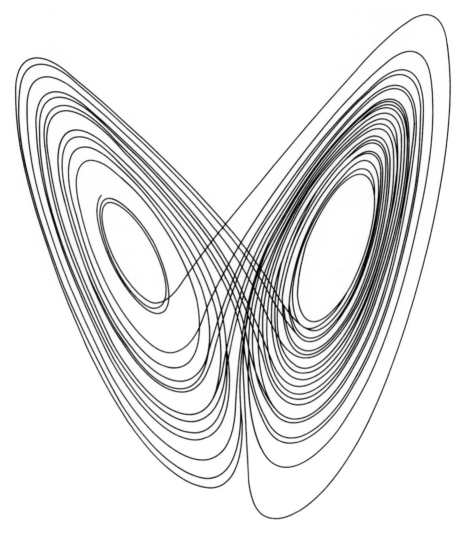

Figure 6.2 A set of nonlinear differential equations modeling air flow in the atmosphere generates a butterfly-shaped figure called the Lorenz attractor.

Further study of both simulated and observed chaotic attractors shows that they are fractals, usually of a dimension greater than two. As we saw in Chapter 5, magnifying a fractal shows more detail, no matter how much magnification is done. Furthermore, chaotic attractors act like amplifiers to bring microscopic fluctuations to macroscopic expression—small-scale uncertainties are made larger and larger. Because the starting point can never be known with sufficient precision, no exact solution—and no mathematical or theoretical shortcut to predicting the future—is possible.

This curious consequence implies that prediction isn't necessarily a good test of a theory. The classical approach to verifying a theory is to make predictions and test them against experimental data. If the phenomena are chaotic, however, long-term predictions are intrinsically impossible. This has to be taken into account in judging the merits of a theory. The process of verifying a theory thus becomes a delicate operation that sometimes relies more on statistical and geometric properties than on detailed prediction.

One important insight that these dynamical studies provide is that systems may behave "normally" for a wide range of initial conditions and then suddenly shift into a chaotic mode when one parameter, such as the concentration of an administered drug in the case of a neural system, exceeds a critical value. Thus, a tiny change in parameter can result in dramatically altered behavior. This abrupt change in behavior, obtained reproducibly in response to a small change in the value of a system parameter, is, in mathematical terms, a bifurcation, or branching.

In the case of epilepsy, researchers are exploring whether transitions from normal behavior to convulsions are the result of bifurcations. Such knowledge would greatly facilitate treatment. A reliable mathematical model would allow physicians to identify which drugs reset parameters so that a convulsion ceases.

Mathematical methods are now available that permit the analysis of dramatically irregular behavior. Highly disordered behavior no longer need be viewed as uncontrollable or inevitable. It can, at least, be analyzed by matching the appropriate mathematical model with the right physical situation.

The Stranger Side of Squares

The penetrating squeal that escapes from a loudspeaker when a microphone picks up the speaker's sound is an ear-splitting example of feedback. The output from the microphone and amplifier is continually cycled back into the system as new input. Rapidly, the whole process gets out of hand.

Mathematics has its own version of feedback, and its results are often equally startling. In this case, the amplifier is an algebraic expression, such as $4x(1 - x)$. Computing the value of that expression for some initial value x, then substituting this answer back into the original expression starts the loop. Repeating this simple, iterative process over and over again leads to surprisingly complex, even unpredictable, mathematical behavior. Each step, though completely determined by the equation, may take a traveler on an unexpectedly erratic course. This mathematical phenomenon expresses the same kind of disorder that arises with nonlinear differential equations. In this instance, however, the equations are exceedingly straightforward.

A simple calculator experiment illustrates what may happen. The sequence of numbers 0.20, 0.64, 0.922, 0.289, 0.822, 0.586, 0.970, 0.116, 0.406, . . . looks haphazard, but it's the result of starting with the number 0.20 and substituting that value for x in the equation $4x(1 - x)$. This gives the second number in the sequence, which is 0.64. Substituting $x = 0.64$ into the same expression gives the next answer, 0.922, and so on.

Further exploration in this astonishingly chaotic regime turns up another surprise. A slightly different starting value leads to a sequence that bears little resemblance to one initiated by a near neighbor. For the seed value 0.21, for example, the sequence is 0.21, 0.664, 0.892, 0.364, 0.926, 0.275, 0.796, 0.650, 0.910, . . . , a far cry from the sequence starting at 0.20.

A calculator experiment does not, however, provide conclusive evidence that something unusual is going on. The results may be affected by rounding off answers to three decimal places, the chosen examples may be special cases, or the iteration may not have been carried out for enough steps. What's needed is a more thorough, systematic look at the same process applied to a wide range of equations. With some equations, nothing much happens. For a linear expression such as $3x$, each iteration simply triples the initial value of x, no matter what the initial value. The numbers in the sequence get larger and larger in predictable steps. Iterating something like x^3 also results in values that keep going up and up or down and down. In fact, after n steps, the nth value of the sequence is easily determined—it equals x^{3^n}. Clearly, to see cyclic or erratic behavior, the mathematical expression must have some sort of switchback or twist.

The simplest example of such a feedback system is represented by the nonlinear equation $x_{n+1} = kx_n(1 - x_n)$. This equation determines the future value of the variable x at time step $n + 1$ from the past value of x at time step n. Known as the logistic equation, it is sometimes used in ecology as a rudimentary model for predicting population growth. The variable x can be thought of as the population of, say, gypsy moths at a particular time. The equation expresses how the population in a given year depends on the population in the previous year. The parameter k can be adjusted to reflect a variety of biological constraints. An enormous amount of work has gone into understanding the behavior of sequences defined by this equation, over a wide range of initial conditions and parameter values.

When the logistic equation is written in the form $x_{n+1} = kx_n - kx_n^2$, its true mathematical identity becomes clearer. This is a quadratic equation with a linear first term and a nonlinear (quadratic, in this case) second term. The relative importance of the two terms depends on the chosen starting point. For example, when $x_0 = 0.01$, x_0^2 is a minuscule 0.0001. Thus, if x_0 is sufficiently small, the nonlinear term is even smaller and barely affects the computation. In this situation, the population rises steadily when k is greater than 1 and falls steadily when k is less than 1. In other words, the linear term can be interpreted as a birth or death rate, leading to exponential growth or decay. On the other hand, when x_0 is large, the nonlinear, or feedback, term becomes important. Because this term is negative, it represents a death rate that dominates when the population gets too large. Biologically, that nonlinear death rate could be the result of food supply shortages, widespread disease, or other factors that become important in an overcrowded environment.

For a given value of the parameter k, once a starting point x_0 is specified, the evolution of the system is fully determined. One step inexorably leads to the next. The whole process can be plotted on a graph to produce a sort of guide map on which the path from one value of x to the next can be traced. Such a graph is called a return map.

For the logistic equation, plotting x_{n+1} against x_n results in a parabola that opens downward like an upside-down bowl. The inverted parabola sits on the x_n axis, wedged between 0 and 1. Its rounded peak reaches a maximum value of $x_{n+1} = k/4$ when $x_n = 0.5$. With this graph, a logistic explorer can track the equation's numerical evolution, as long as the starting point is between 0 and 1 on the x_n axis.

The simplest course is to locate a starting point x_0 on the x_n axis, then shoot vertically upward to intersect the parabola. A horizontal glance from the parabola toward the x_{n+1} axis provides the value x_1. Returning with x_1 as a new starting point on the horizontal (x_n) axis allows the whole process to be repeated to obtain x_2.

This graphical procedure seems to be almost as much trouble as computing each of the numbers in the sequence, but a handy shortcut simplifies the process. All it takes is the addition of a slanted line 45 degrees up from the horizontal axis (representing the line $x_{n+1} = x_n$). Now the best course to steer is from x_0 vertically to the parabola to reach x_1, then horizontally to the 45-degree line and vertically back to the parabola for x_2, and so on. Each step, or iteration, requires a horizontal move to the 45-degree line, then a vertical jump or drop to the parabola (see Figure 6.3).

These paths, or orbits, give the first indication of which routes lead to the erratic behavior of chaos. Whereas some orbits converge on one particular value, others jump back and forth among a few possible values, and many roam, never settling anywhere. Exactly what happens depends strongly on the value of the parameter k, which measures the degree of nonlinearity. As k increases, going from 0 to 4, the logistic equation shows an amazing transformation from order to chaos.

The transformation happens in clearly defined stages. When k is less than 1, every starting point produces a path that eventually ends up at 0. When k is between 1 and 3, just about every route, no matter where it

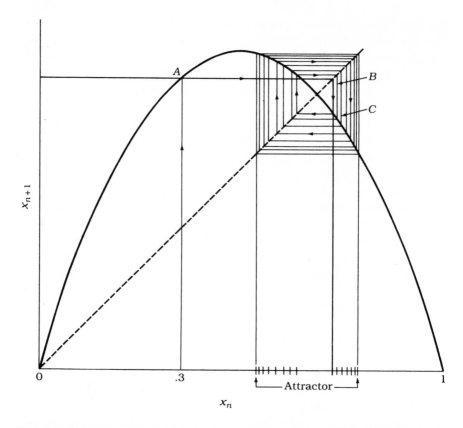

Figure 6.3 The logistic map defines an inverted parabola. Once the initial value is specified, values of x at succeeding time steps can be found by tracing a path vertically up to the parabola at point A, horizontally to point B on the 45-degree line, vertically down to point C on the parabola, and so on. Eventually, the path settles down into a cycle, going back and forth between two values, which comprise the attractor.

starts, is eventually attracted to a specific value called a fixed point, which occurs where the parabola intersects the 45-degree line at $x = (k - 1)/k$. This corresponds to a steady-state, or equilibrium, population, which occurs when the number of births and deaths balance.

When k is larger than 3, the fixed point becomes unstable, and wild and woolly things begin to occur. At $k = 3.2$, two fixed points appear, and the value of x oscillates between the two values of x_n, roughly equal to 0.5 and to 0.8. In this case, the orbit for any starting point is trapped in a period-2 cycle, that is, it skips relentlessly from one to the other value. Boosting the value of k adds two more fixed points, and the orbit visits all four values in turn, returning to the original point in four time steps. That behavior puts these orbits in a period-4 cycle. As k grows, the period and number of fixed points double and redouble, until at k approximately equal to 3.57, both the period and the number of fixed points are essentially infinite. The trajectories for practically all starting points appear in the sequence of successive values of x. The abrupt population changes evident for these values of k are indistinguishable from a random process, although there are no random forces involved, and, according to the logistic equation, the future is strictly determined by the starting point x_0.

The cascading period-doubling route to chaos is itself worth exploring. The range over which a particular cycle is stable decreases quickly as the cycle's period and the number of fixed points increase. Longer and longer periods are crowded into less and less space. In fact, the physicist Mitchell J. Feigenbaum, having observed this period-doubling sequence in numerical experiments on a computer, was able to prove that the successive stable intervals become smaller in a very particular way. The range of values of k over which the period-2 cycle is stable is about 4.6692016 . . . times wider than the range of the period-4 cycle; the range for the period-4 cycle is the same factor wider than the period-8 cycle, and so on, leading to the rapid accumulation of cycles with longer and longer periods until the limit at infinity is reached.

To get a better idea of what is happening as periods increase and chaos begins to take over, we need another graph. Plotting successive values of x on one axis against the parameter k on the other, as k changes from 3.5 to 4, provides a sufficiently broad canvas against which the logistic equation's strange behavior can be viewed. Each return map for a particular value of k, iterated hundreds of times, is compressed into a narrow vertical strip. Piecing together these strips in a wide band produces the computer-generated mural known as a bifurcation diagram (*see Figure 6.4*).

The result is a graphic display of the underlying structure of chaos. Forking patterns show an orderly progression from regular to chaotic behavior. Wherever the long-term behavior of orbits for a give value of k converge to a cycle of, say, period 4, the diagram shows four discrete values of x. Indeed, for $k = 3.5$, the system eventually settles down to a periodic oscillation among four values. Where the evolution is chaotic, the values of x seem to cover the whole interval, leaving a dark strip. However, even the chaotic domain is itself crisscrossed by dark streaks, showing that some x values are visited more often than others. Intersecting streaks signal mathematical crises, where nearby orbits are forced to wander in a tight space. A particularly dramatic collision is visible at $k = 3.68$.

Windows of periodic behavior also sit embedded in the chaotic regime. A period-3 cycle, for example, shows up at k roughly equal to 3.83. Here the population increases in two successive years and decreases in the third. As k edges upward, that cycle becomes unstable, the period doubles, and the number of fixed points doubles to six, then to 12, then 24. It turns out that

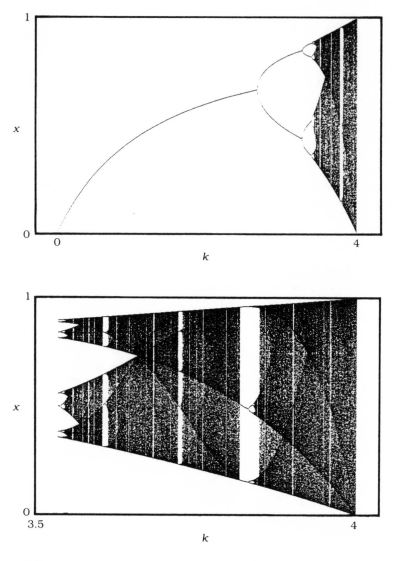

Figure 6.4 *Top:* The bifurcation diagram for the logistic map. *Bottom:* A detailed look at the region between $k = 3.5$ and $k = 4.0$, when chaos makes its appearance.

there are windows of stability for cycles of all whole-number periods, each of which exhibits a cascade of period-doubling bifurcations back to chaos. Yet despite the infinite number of such intervals of stability throughout the range of k, there's still plenty of room for truly chaotic motion. In other words, if the exact evolution of x_n looks chaotic, then it probably is. It's unlikely to be evidence of very long but periodic cycles. By the time $k = 4$—

and x can end up anywhere between 0 and 1—it's relatively easy to show that the orbits meet the definitions of both a chaotic and a random process.

At first glance, the complexity of the dynamics shown by even a simple expression such as the logistic equation is somewhat dismaying. It's like stepping into a modest cottage and finding a veritable palace inside, full of secret passages and mysterious rooms. What happens with more complicated equations? Luckily, features like the Feigenbaum constant and period-doubling bifurcation sequences appear in a variety of systems (*see Color Plate* 9). Whole classes of equations, when iterated, exhibit exactly the same behavior as quadratic polynomials. These features are universal, in the sense that they depend solely on the presence of feedback and are virtually insensitive to other system details.

Feigenbaum's period-doubling route to chaos has also been observed in physical systems as diverse as turbulent flows, oscillating chemical reactions, nonlinear electric circuits, and ring lasers. The logistic equation itself seems to mirror well the wild and unpredictable fluctuations that may occur in gypsy moth populations, in stock and commodity prices, and in the dynamics of some mechanical oscillators.

——— — — —-- The Computer ——— —— ——— —— as Microscope

There are few more striking demonstrations of the complexity hidden in simple laws or rules than the multitude of structures contained within a mathematical object called the Mandelbrot set. This figure is visually one of the most complicated objects in mathematics. From within its bounds and around its borders come pictures of order and chaos, conflict and coexistence, stasis and transition. Enigmatically beautiful and astonishingly intricate, such mathematical products of computer graphics also represent attempts to understand how subtle changes in equations affect their behavior.

The polynomial $x^2 - 1$, under iteration, does nothing particularly exciting. From any starting point defined by an ordinary decimal (or real) number, successive values show a humdrum predictability. The fireworks begin when complex numbers are used in place of real numbers. With complex numbers, what was only a crack in a wall becomes a full-fledged picture window, revealing a truly wondrous, chaotic landscape.

Real numbers can be thought of as labels of all the points on a line. For example, the real number 1.5 lies halfway between points 1 and 2, and 1.67 lies at a specific location between 1.5 and 2. Every real number has its place along the number line, and every point on the number line has its own real-number label. Complex numbers, on the other hand, define a point not on a line but somewhere in a vast two-dimensional sheet of numbers called the complex plane.

A complex number, z, consists of two parts, which for historical reasons are called real and imaginary. These terms no longer have a literal meaning. The two parts could just as easily be named Tweedledum and Tweedledee or Albert and Brenda. A complex number may be written in the form $a + bi$, where a is the real part and bi is the imaginary part. The italic i next to the b shows which part of the complex number is imaginary. In this notation, $4 + 3i$, $0.5 + 1.76i$, and $16 - 8i$ are all examples of complex numbers.

The two parts of a complex number can be pictured as the coordinates of a point plotted on a plane. Locating a particular complex number requires defining two axes: a horizontal real axis and a vertical imaginary axis. Assuming the axes intersect at the point $0 + 0i$, then the number $4 + 3i$, for example, would lie 4 units to the right and 3 units vertically upward from the origin where the axes intersect. This point would be 5 units away from the origin. The complex plane is filled with uncountably many such pairs of numbers. Their real and imaginary parts can be either positive or negative and either whole numbers or decimals.

Adding two complex numbers is easy. To find the sum of $4 + 3i$ and $2 - 7i$, the real and imaginary parts are combined separately. In this case, the sum is $6 - 4i$. Under addition, the two types of terms stay segregated.

Multiplying two complex numbers is only slightly more difficult. If the symbol i is treated like an x in algebra, the product of $4 + 3i$ and $2 - 6i$ is $8 - 24i + 6i - 18i^2$. It happens that $i^2 = -1$. The product now becomes $8 - 24i + 6i + 18$. Simplifying by collecting the real and imaginary parts reveals the final answer: $26 - 18i$.

With rules for addition and multiplication and a vast plane to explore, a new journey in search of chaos and complexity can begin. The same questions that come up for expressions involving real numbers also arise for expressions involving complex numbers. Will the iterative process, starting at an arbitrarily chosen value, generate a sequence of numbers that gets closer and closer to a particular value and eventually comes to rest there? Will the sequence arrive at a cycle of values, repeated over and over again? Or is the sequence erratic and unpredictable?

For the expression $z^2 - 1$, some choices of starting values generate sequences that grow without bound, flying off to infinity. Other starting values lead to sequences that stay within a well-defined boundary on the plane. This boundary has the fractal structure reminiscent of a deformed island coastline spotted with wartlike protuberances and pinched into deep bays (*see* Figure 6.5). Magnifying a small piece of the border reveals details that mimic the overall raggedness of the entire border.

This kind of boundary is known as a Julia set, named for the French mathematician Gaston Julia (1893–1978), who, along with Pierre Fatou

Figure 6.5 Iterating the expression $z^2 - 1$ generates a Julia set, in which the black area represents the set of complex-number starting points that remain bounded.

(1879–1929), first studied these forms in the early part of the twentieth century. They established the idea that the entire boundary could be re-generated from an arbitrarily small piece of the boundary. Their fascinating work, however, remained largely unknown, even to most mathematicians, because Julia and Fatou lacked the tools needed to communicate their subtle ideas. High-resolution computer graphics now brings their visions to life.

Different quadratic equations lead to different pictures of the behavior of sequences. One particular set of equations can be written as $z^2 + c$, where c is a given complex number. Any computer can perform the necessary iteration: Square the starting number, add the constant to generate a new number to be squared, and so on, over and over again. Because complex numbers represent the coordinates of points in two dimensions, each iteration can be viewed as a hop from one point in the plane to another. As in the case of real numbers, the set of hops can be thought of as a trajectory or an orbit.

When $c = 0$, three types of behavior are possible. All starting points that lie within a distance of 1 of the origin are drawn toward the origin. It's as if a magnet of strictly limited range was sitting at the point $0 + 0i$, attracting any hopping particles that happen to start within its circle of influence. In fact, zero is said to be an attractor for the process. Successive squarings yield orbits that tend to zero.

All points more than a distance 1 from the origin end up sailing farther and farther away. Somewhere at infinity, another magnetic attractor is doing its job. As the iteration continues, the numbers steadily become larger.

Points on the boundary, a circle of radius 1, stay where they are, caught in the competition between the two attractors. Overall, two zones of influence divide up the plane, and the boundary between them is simply a circle. This boundary is an especially simple Julia set.

Varying c leads to an infinite number of different pictures. For $c = -0.12375 + 0.56508i$, the central attractor is no longer zero, and its outer boundary—the limit of its circle of influence—is no longer smooth but crumpled. When $c = 0 + 1i$ (or just plain i), an orbit starting at 0 is drawn into a never-ending oscillation between the two complex numbers $-1 + 1i$ and $0 - 1i$. For other starting points, the sequence may also be attracted to two fixed points, or it may explode to larger and larger values. Picking $c = -0.12 + 0.74i$ yields not a single island of attraction but an infinite number of barely connected, deformed circles.

Clearly, by varying c, an incredible variety of Julia sets can be generated. Some look like fat clouds; others are like twisted, thorny bushes; and many look like sparks floating in the air after a flare has gone off. With a little imagination, a person flipping through a book of these Julia sets may glimpse the shape of a rabbit, the jagged form of a dragon, and the curls of a sea horse. Sometimes, the Julia set is merely a dust of disconnected points (*see Figure 6.6*).

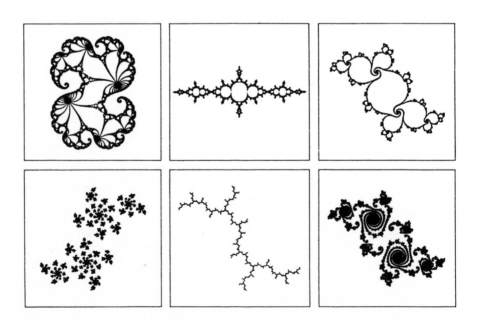

Figure 6.6 Julia sets when iterated for various values of c in the expression $z^2 + c$ come in many forms, some connected and some fragmented.

Indeed, the book of Julia sets for iterations of the equation $z^2 + c$ has an infinite number of pages, one for each possible value of c—which covers every point in the complex plane! When mathematicians began to study all these forms, they soon realized that they needed a kind of table of contents that would give them an overview of this huge book filled with page after page of incredibly complex figures. The organizing principle was discovered in 1980 by Benoit Mandelbrot, and it now carries his name—the Mandelbrot set.

In every case, depending on their starting points, some orbits stay bounded no matter how long their hops are tracked, whereas others streak off to infinity. This difference in behavior makes the boundary between these two regions of crucial importance in defining the properties of a given equation.

It turns out there are only two major types of Julia sets, no matter what c is. Either the area within a boundary is all one piece (that is, the area is a connected structure), or it's broken into an infinite number of separate pieces to form a cloud of points, called a Cantor set and discussed in Chapter 5. No matter where anyone looks in the book of Julia sets, each page falls into either the first or the second category. The Mandelbrot set consists of all values of c that have connected Julia sets.

That definition makes it possible to draw a portrait of the Mandelbrot set. If the Julia set corresponding to a certain complex number c is a cloud of dust, that point in the complex plane is colored white. If the Julia set is connected, the point is colored black.

If the only way to draw the Mandelbrot figure meant looking at all the different Julia sets and deciding which ones belong and which ones don't, it would take an eternity to compile the table of contents. However, the mathematicians John Hubbard and Adrien Douady found a quick way to generate the set. They proved that it's enough to know for a given value of c whether the starting point 0 stays bounded. If it does, the point c belongs to the Mandelbrot set. Hubbard describes his procedure, evaluated for all values of c, as probably the most economical definition yet known of a complex object. A computer program written to generate the Mandelbrot set need be no more than four or five lines long.

Finding the set comes down to computing the sequence of numbers: c, $c^2 + c$, $(c^2 + c)^2 + c$, . . . and determining whether the numbers resulting from a particular choice for the initial value c steadily grow larger or instead stay bounded, never rising above a certain value. The Mandelbrot set consists of all choices of c that stay bounded. For example, $c = -1 + 0i$ (or -1) is in the Mandelbrot set because its sequence hops back and forth between 0 and -1. The point $1 + 0i$, on the other hand, goes from 0 to 1 to 2 to 5, continuing ever upward, fleeing toward infinity.

The rotund, wart-covered, black figure that emerges from evaluating Hubbard's sequence is, in effect, the required one-page table of contents

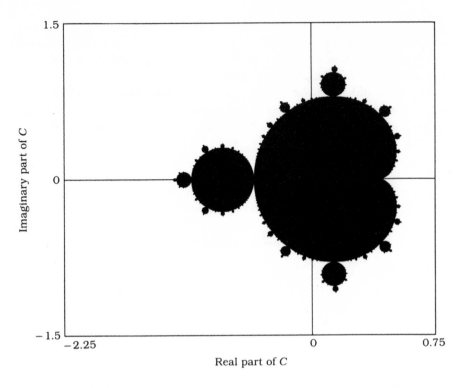

Figure 6.7 The Mandelbrot set (*shown in black*) extends from the cusp of the cardioid at $c = 0.25$ to the tip of its tail at $c = -2$.

(*see Figure* 6.7). This awesome image says something about all quadratic equations at once, providing precise information about every page in the book of iterated squares.

The figure itself, looking like a pimply, squat snowman lying on its side, takes up only a small piece of the entire complex plane. Its pointed cap reaches -2 along the real axis, while its indented bottom rests at $\frac{1}{4}$. Its arms extend out along the imaginary axis to i and $-i$.

This strange, solitary figure is a source of unending wonder. Mathematicians such as Hubbard, Heinz-Otto Peitgen, and John Milnor have been unable to resist its allure and have expended a great deal of computer time exploring the figure's intricacies, suggesting and proving mathematical conjectures in the course of their computer-aided explorations. They have also constructed a considerable library of magnificent graphic images, portraits of the Mandelbrot set in all its moods.

The Mandelbrot set has detailed structure on all scales. Zooming in for a closer look shows fine structures that seem to get more and more fantastic and complicated as the magnification increases. Close-ups of its borders

Figure 6.8 A blowup of a piece of the Mandelbrot set's boundary reveals a riot of filaments and small copies of the Mandelbrot set.

unveil a riot of tendrils and curlicues, yet everything is connected. A bewildering array of delicate filaments holds all the parts together (*see Figure* 6.8).

Delving deeper and deeper also turns up miniature snowmen. Little Mandelbrot figures lurk wherever the computer peeks. For example, the Mandelbrot set wears a coat of fine, crumpled, branched antennae. A close inspection reveals that these antennae also carry many little copies of the larger Mandelbrot set, sitting like fuzzy, round marshmallows strung out on a skewer. Further magnification turns up even tinier copies of the Mandelbrot set. In fact, the Mandelbrot set includes an infinite number of copies of itself, all tied to one another by tiny lifelines. The details are always different, however, depending on the magnification and the figure's precise location.

Remarkably, each point in the Mandelbrot set carries its own address. It's possible to figure out where a blown-up region fits just by counting the number of strands that come together at various nearby points. Moreover, the Mandelbrot set does much more than an ordinary table of contents. It not only catalogs where everything is but also contains a miniature of each of the book's pages in an incredibly compressed form. Looking at a close-up of a tiny piece of the Mandelbrot set provides a glimpse of the Julia sets associated with those values of c (see Figure 6.9).

If c happens to lie well within the main body of the Mandelbrot set, the Julia set is a crinkled circle, surrounding a single attractor. If c is inside one of the snowman's buds, the Julia set consists of infinitely many fractally deformed circles. Picking a value of c from one of the set's branched antennae reveals a similarly furry Julia set. Every neighborhood has its own distinctive character.

Following a path that starts deep within the Mandelbrot set then crosses its boundary reveals a dramatic fundamental change in the nature of the associated Julia sets. At the very edge of the Mandelbrot set, the Julia sets explode, splintering and falling to dust. As the path streaks into distant parts of the complex plane, far away from the Mandelbrot figure, the scenery gets more sparse. The clouds of points in the Julia sets get thinner and thinner.

The greatest diversity flourishes in the border zone, where many cultures and customs are apt to mingle. This tangled region harbors a fantastic, baroque coterie of dragons, sea horses, and other strange creatures, contrasting starkly with the simplicity of the single mathematical expression responsible for the myriad forms that live in the zone. Attractors compete for influence on the plane. Every conflict breaks up into innumerable smaller battles.

The true complexity of the Mandelbrot set's border region is hard to grasp without the use of computer graphics. A computer can create a colored halo around the figure, making the boundary visible, by tracking how rapidly the sequences of numbers resulting from different starting points grow in size. Starting points that lie within the Mandelbrot set produce sequences of numbers that stay bounded, whereas points outside the figure escape to infinity at varying rates. Each point is colored according to a chart stored in the computer. The gradation of colors indicates how quickly escape occurs, quantifying the escape toward an attractor located an infinite distance away (see Color Plate 10).

The Mandelbrot set serves as a prime example of how simple mathematical operations can yield astonishingly complex geometric forms. The border of this intricate shape twists itself into such a riot of loops and curlicues that it calls to mind the extravagant ornamentation of baroque designs. In 1991, Mitsuhiro Shishikura of the Tokyo Institute of Technology proved that the Mandelbrot set's boundary is as convoluted as the boundary of a two-dimensional object laid out on a flat surface can ever get. In

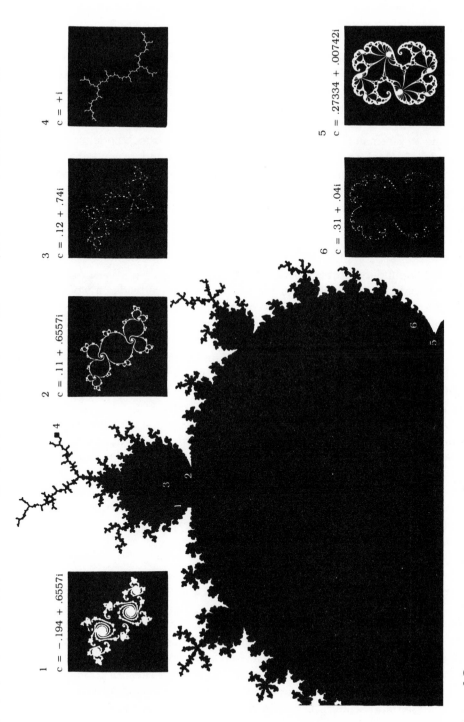

Figure 6.9 The Mandelbrot set determines the structure of corresponding Julia sets. Points that fall within the Mandelbrot set represent connected Julia sets, whereas those outside the Mandelbrot set are matched with fragmented Julia sets.

technical terms, this means that the Mandelbrot set's boundary has a fractal, or Hausdorff, dimension of two.

Hence, the Mandelbrot set's boundary is so convoluted that it appears to have the same dimension as a two-dimensional area even though it is mathematically a curve—albeit an incredibly wiggly one—and curves don't ordinarily have an area. Shishikura also proved that, excluding certain points on the boundary, the boundary's two-dimensional area itself is zero. This means that the boundary is as thick as it can possibly be without truly having an area.

One key remaining question concerns whether the Mandelbrot set's boundary is locally connected. If it isn't, then the boundary would have the same characteristics as a figure in the shape of a spoked wheel, in which any small spotlighted area that doesn't include the hub or rim shows no link between the spokes. Mathematicians have taken several important steps toward resolving the question. If the Mandelbrot set turns out to be locally connected, they can then finally say that they understand the Mandelbrot set completely. If it isn't, the Mandelbrot set contains within its tortuous boundary an even deeper mystery.

The incredible complexity of the Mandelbrot set provides a wonderful world for exploration, with beauty spots to visit, innumerable strange sights to view, and unexpected turns to any journey. In place of the Taj Mahal or the Eiffel Tower, the Mandelbrot tourist visits the unnamed wonders marked by coordinates in the complex plane. Explore the area where the real part lies between 0.26 and 0.27 and the imaginary part between 0 and 0.01, the tour guide suggests. Or why not try -0.76 to -0.74 and $0.01i$ to $0.03i$ or -1.26 to -1.24 and $0.01i$ to $0.03i$.

At the same time, the discovery of the Mandelbrot set is somewhat disquieting. If a mathematical expression as innocuous as a quadratic equation, iterated in the complex plane, exhibits such complexity, what hope is there of understanding more complicated equations? Fortunately, the Mandelbrot set, in various forms, shows up again and again in more complicated systems (*see Color Plate* 11). Little by little, mathematicians are building a new picture of iterated mathematical expressions.

Zeroing in on Chaos

Engineers and scientists spend a lot of time solving equations. Such exercises provide the information needed to understand how molecules stick together, to determine the strength a steel framework must have to support a given structure, to ascertain the size of distant galaxies, or to make any sort of scientific prediction. Equations are the engines that drive much of engineering and science. Over the centuries, mathematicians have developed a variety of methods of solving equations. Using the capabilities of modern computers, they are now starting to explore in detail how the age-old methods work—when the methods can be relied upon, when they fail, and when they behave strangely.

One calculus-based, equation-solving technique that has undergone intense scrutiny is Newton's method, which has been in use for centuries and is often taught in beginning calculus courses. Mathematicians have probed this method's idiosyncrasies and have come up with startling pictures of its behavior. Their results reveal a chaotic side to Newton's method that has become apparent only in the past two decades. Earlier mathematicians had suspected that problems with the method could arise under certain circumstances, but they lacked a way to describe their concerns in a visual manner.

At the center of Newton's method (and much of calculus) is the concept of a function. A function is a rule that assigns a certain output to a given input. The squaring function is a familiar example: input 2, output 4; input 9, output, 81. The reciprocal function turns 2 into ½ and 5 into ⅕. Other well-known functions include the sine and exponential functions.

More formally, a function's output can be written in mathematical shorthand as $f(x)$. Thus, the squaring function can be expressed as $f(x) = x^2$. When the input, x, is 2, the output, $f(x)$, is 4. When $f(x)$ is given as $x^2 + x - 6$, then $f(x)$ has the value -4 if $x = 1$. The process can also be reversed. Sometimes, the value of the function is known and the input value or values are unknown. That requires solving the equation; that is, finding values of x that make the equation true.

Anyone who has studied high-school mathematics is probably familiar with the problem of finding values of x for which, say, $x^2 + x - 6 = 0$. In this case, x can be either 2 or -3. The two solutions, 2 and -3, are known as the roots of the equation.

One way to picture what is happening is to plot a graph that shows what the function $x^2 + x - 6 = 0$ looks like. When the function's value, $f(x)$, is computed for various values of x, the resulting pairs of numbers represent coordinates that can be plotted on a graph. For example, when $x = 1$, $f(x) = -4$. The point $(1, -4)$ would lie one step away from the graph's vertical axis and four steps below the horizontal, or x, axis. The resulting curve is a parabola that happens to cross the x axis twice, at $x = 2$ and $x = -3$ (see Figure 6.10).

Solving the equation $x^2 - 5 = 0$ is a little trickier. There are two answers: plus or minus the square root of 5 ($\pm\sqrt{5}$). But what is the numerical value of $\sqrt{5}$? In other words, exactly where does the curve represented by this equation intersect the x axis?

Newton's method provides one way of finding where an equation crosses the x axis or, more generally, the roots of an equation. The procedure begins with a guess. In the case of $\sqrt{5}$, the solver may start with 2 as a trial value of x. Applying the formula that lies at the heart of Newton's

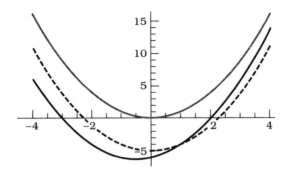

Figure 6.10 Plots of the quadratic functions x^2 (*thick line*), $x^2 + x - 6$ (*solid line*), and $x^2 - 5$ (*dashed line*) showing where the functions intersect the x axis.

method produces a new, improved estimate of the root—2.25. The process is repeated using the improved estimate as the input, and the result, 2.236, is an even better estimate. This iterative procedure continues until the solver is satisfied with the answer's accuracy.

To picture what is happening requires a look at the graph of a typical function—a smooth, undulating curve that happens to cross the x axis at some point (*see Figure* 6.11). The first guess locates a point on the curve. Newton's method defines a tangent, the unique line that passes through a

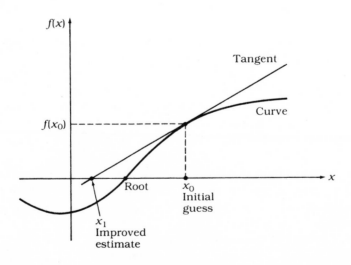

Figure 6.11 Newton's method can be used to find a root or where a function crosses the x axis.

9 Branching to Infinity

10 Culture Clash

11 The Ubiquitous Snowman

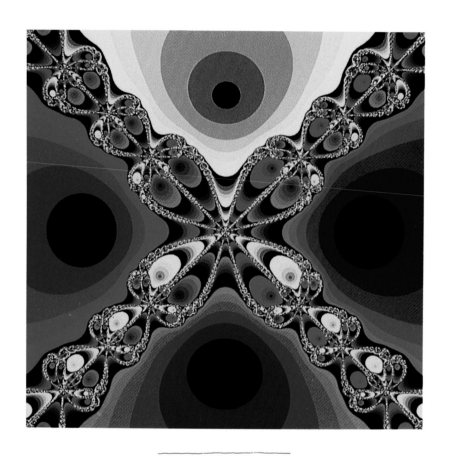

12 Basins of Attraction

13 Bursting into Chaos

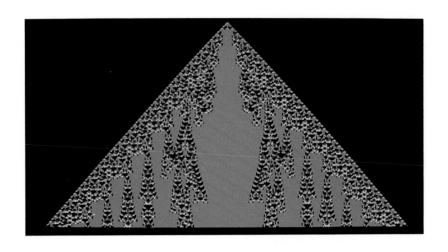

14 Grid Rules

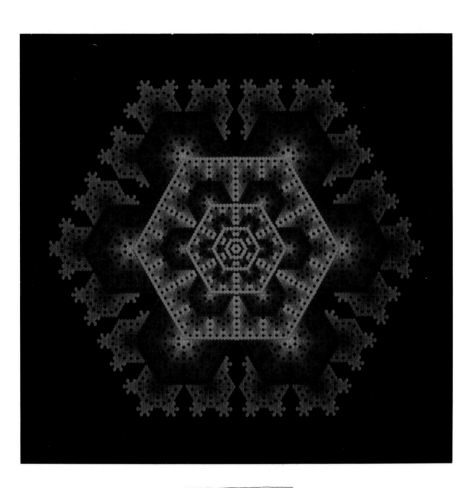

15 Snowflake Cells

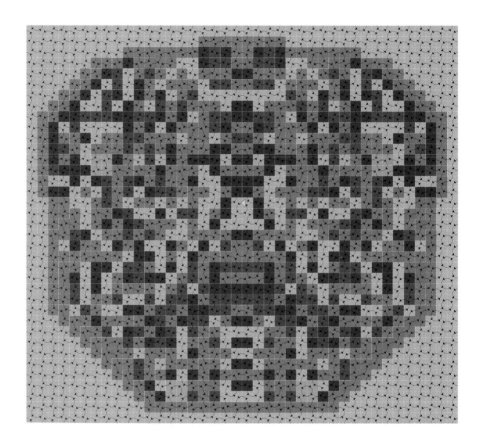

16 Ant Tracks

given point and represents the slope of the curve at that point. Extending the tangent until it crosses the x axis provides a new value of x, which is presumably a better estimate of where the original curve crosses the x axis. The process can then be repeated by substituting this new value of x for the original guess, and so on.

Such a procedure can be converted easily into a reasonably efficient computer algorithm for finding one or more of the roots of practically any polynomial equation. The problem with using the method, however, turns out to be one of selecting an appropriate starting point, or making the right initial guess. Not all choices of starting point converge quickly on a specific root.

The potential problems become much more serious when Newton's method is applied to functions of a complex variable. Equations involving complex numbers bring in two dimensions. As described earlier, each complex variable, z, can be thought of as having two components; whereas real numbers can be represented as points on a line, complex numbers must be located on a plane.

When complex numbers were invented centuries ago, no one could think of any practical uses for them. Now, they regularly show up in methods for solving differential equations and in other applications of calculus. They also play an important role in describing physical phenomena, such as electromagnetism and the properties of electric circuits.

A polynomial equation such as $z^4 - 1 = 0$ has as many roots as the highest power to which z is raised in the equation. In this instance, there are four complex roots. The equation $z^{17} - z^5 + 6 = 0$ has 17 solutions, or roots. In terms of Newton's method, these roots act as centers of an attractive force field that draws a sequence of guesses closer and closer to one of the roots.

When Newton's method is used to find a specific root, the solver hopes that the chosen starting point leads quickly to the appropriate root. However, the method fails when the chosen point happens to fall on a boundary separating regions "controlled" by different roots. Failure may also strike when, for some starting values, the procedure gets sidetracked and, like a needle stuck on a phonograph record, ends up oscillating back and forth between two numbers, neither of which is a root.

Computer graphics provides a way to picture what is happening. For a given equation, the computer applies Newton's method for each value of z—perhaps a million or more different values. For each starting value, the computer determines toward which root that value converges and assigns a color to the point. Shades of the color indicate how quickly that point comes close to the root.

The result is a glowing tapestry that often features large pools of color. These basins of attraction, as they are known, are "safe" areas. Any starting point selected from a safe region, within a reasonable number of iterations, comes close to a root. The equation $z^4 - 1 = 0$ has four such basins (*see Color Plate* 12).

Life near or at a boundary, however, is considerably more complicated. The borders, much more than simple dividing lines, consist of elaborate swirls and whirlpools that can pull Newton's method into any one of the four roots. In these vicinities, a minuscule shift in starting point can lead to widely divergent results. And right on the convoluted boundary itself lie points that lead to no root. These points comprise the Julia set.

There's a further complication. Magnifying a section of the boundary region doesn't simplify the border. Instead, the magnified section looks a lot

Figure 6.12 When Newton's method is applied to the equation $z^4 - 1 = 0$, the resulting Julia set has four basins of attraction. As this close-up shows, the boundary between two such basins may be extremely complicated.

like the original picture of the boundary. Further magnification merely reveals more miniature copies of the overall structure. Like a fractal, it is self-similar (*see Figure* 6.12).

Mathematicians are also interested in which characteristics of a function lead to particularly chaotic or unpredictable behavior. To study this problem, they pick a family of functions and vary one parameter to see what happens. One famous example, first probed in 1879 by the British mathematician Arthur Cayley (1821–1895), is the family of cubic polynomials. Cayley eventually had to give up his search because he found the answer too complicated. It took computer graphics to illustrate what was happening. The expression $z^3 - \alpha z + \alpha - 1$ represents a one-parameter family of cubic polynomials. Every polynomial in this family has a root (or zero) at $z = 1$. The location of the other two roots depends on the numerical value of the parameter α. By varying the value of α, which can be either a real or a complex number, mathematicians can explore how the pictures associated with each function change. That exploration involves iterating the expression $[2z^3 - (\alpha - 1)]/(3z^2 - \alpha)$.

The results are visually spectacular. At the boundaries between the basins of attraction, all three basins (and all three colors) must meet at every point (*see Figure* 6.13). That leads to an intricate interweaving of color. Changing the parameter α shifts the boundaries and ruffles the basins of attraction. For certain values of the parameter, Newton's method may take a point into a never-ending oscillation between two values, interrupting its quest for a root. In that case, the point falls into an attracting cycle. In some other cases, the basin of attraction around a zero practically disappears, and no choice of points leads to the root. The zeros themselves become unstable. The disappearance of basins of attraction marks the onset of chaos.

Cubic polynomials are not the only functions worth exploring. Algebraic expressions such as $(z + \alpha)^{\alpha(z - 1)}$ lead to equally startling and complicated patterns when Newton's method is applied. The resulting pictures reveal intricate basins of attraction, evidence of bifurcations, and hints of chaos. The snowmanlike Mandelbrot set appears from time to time. The Julia-set portraits of various functions show important differences as well as striking similarities.

Studying the descent of Newton's method into chaos is more than just generating pretty pictures. Ultimately, what mathematicians and scientists want is a good method for solving equations. Computer experiments, conjectures suggested by the observations, and new theorems are leading to a deeper understanding of Newton's method and its limitations. Mathematicians are gradually building a fence around this venerable mathematical technique for solving equations—a fence to show where it's safe to tread and which territories to avoid.

Figure 6.13 The bull's-eye patterns indicate areas in which starting points for Newton's method converge toward one of the three roots of the equation $z^3 - 1 = 0$. The root itself falls within the small circle near the upper left-hand corner. The boundary region is a fractal because the overall pattern repeats itself on smaller and smaller scales, as seen in the teardrop shapes scattered throughout the picture.

Sudden Bursts and Quick Escapes

A plume of smoke rises lazily into still air. Drifting upward, it shows a smooth profile against the sky. When the smoke column gets high enough, however, it begins to drift, first edging to one side, then swinging back.

Soon the plume no longer has a well-defined form. It breaks up into a disorderly tangle of wisps.

The transition from a smooth to a turbulent flow can be seen not only in smoke plumes but also in water squirting from a garden hose, in air flowing over an airplane wing, and in curling ocean waves as they break into foamy whitecaps. Similar shifts from orderly to disorderly behavior show up in populations of creatures such as gypsy moths or in the erratic chemical changes that disrupt oscillating chemical reactions. In each case, predictable behavior gradually degenerates into chaos. In well-defined stages, the systems become increasingly irregular. Simple mathematical equations, such as the logistic equation or an iterated quadratic polynomial, often fit this type of pattern.

The shift into chaos is extremely abrupt, however, in the violent contortions of flames during combustion or in pockets of agitated air that can suddenly send an airplane crashing to the ground. Such systems practically explode into chaos. Under the right circumstances, a class of simple mathematical expressions also bursts into chaos. Known as transcendental functions, they include the exponential, sine, and cosine functions. The exponential function, represented by e to some power x, is familiar to anyone who has dealt with compounded growth, whether in populations or in accumulated interest in a savings account at a bank. The sine and cosine functions (usually written as sin and cos), often associated with angles, come up in numerous trigonometric applications, such as navigation and surveying, and for describing periodic phenomena, such as waves and alternating currents.

On the computer screen, iterations of these functions manifest themselves as dazzling images of iridescent dragons clawing their own tails, swirling, rainbow-hued galaxies scattering vivid sparks, and geometric jets spewing colored streams into still black basins (*see Color Plate* 13). Moreover, by varying a single parameter, these iterated functions may burst into complete irregularity once a certain threshold is crossed.

Robert Devaney of Boston University, who has explored these functions in their complex world, stated: "Admittedly, these dynamical systems are approximate models of real physical systems, but the burst illustrated with these simple models holds the key to understanding similar phenomena in more complicated settings." Devaney's technique for studying the dynamic behavior of transcendental functions is equivalent to entering a number into the display of a scientific calculator, locating the exponential, sine, or cosine key, then repeatedly pressing the appropriate button and observing what happens to successive numbers displayed. If the calculator is able to deal with complex numbers, it becomes possible to wander across a broad plane, not just along a narrow road.

Each iteration represents a step along a path that loops from one complex number to the next. The collection of all such points along a path constitutes an orbit. The basic goal is to understand the ultimate fate of all orbits of a given system. Depending on the value of z chosen as a starting point, the orbit sometimes behaves tamely. The same answer may come up every time (a fixed point), or the answers may stay close to the original value or even return to the original value after a certain number of iterations (a periodic point). On the other hand, the answers could get steadily larger.

An orbit is considered to be stable if neighboring starting points have orbits that behave similarly; the orbits stay roughly in step. Less predictable, or chaotic, orbits have nearby points that quickly diverge. Changing the starting point z_0 ever so slightly may generate a vastly different orbit.

The starting points of chaotic orbits can be color-coded to indicate how quickly the points escape along their orbits to infinity. In Devaney's computer-generated maps of chaotic regions, red represents points that explode beyond a certain value in only one or two steps. The colors orange, yellow, green, blue, and violet represent successively slower rates. Black areas encompass points that, upon iteration, map into values that do not escape. The black areas, called basins of attraction, are stable regions. All points that are colored black, under iteration, tend toward fixed points or toward periodic points called attractors. The colored areas represent unstable, chaotic regions. For values of z in the chaotic regions, the chosen function seems to behave randomly.

The colored regions of a given complex function also give the barest outline of a Julia set, defined a little differently here than in the case of polynomial equations, where the Julia set is simply the boundary between attractors. In this case, the Julia set contains all points that seem to drive neighboring points farther and farther away. In other words, these orbits are completely chaotic. The collection of these special points corresponds to a strange repeller. The complex plane just divides into two intricately shaped regions—basins of attraction centered on attractors and Julia sets corresponding to strange repellers.

The Julia sets that Devaney finds are often fractals as well. By examining any of the patterns closely, one finds that their features tend to replicate themselves on smaller and smaller scales. A fist bursts into fingers that each burst into smaller fingers, and so on.

A calculator experiment, using real numbers, gives a hint of how the parameter k affects the exponential function ke^x, with k always greater than zero. When k is less that $1/e$, the function has two fixed points. Only orbits that start at real numbers greater than the largest of the fixed points escape to infinity. However, when k is greater than $1/e$, all orbits head for infinity, no matter where they start. This behavior may be easily checked with a calculator by iterating ke^x for various choices of the initial value x. It follows that this dynamical system has, for real numbers, two vastly different behaviors. In the complex plane, such behavior is seen as a burst into chaos.

Thus, small changes in a function can radically change the form of the resulting graphs or pictures. Multiplying the exponential by $1/e$ and then iterating the function over a grid of complex-numbered starting points results in a picture that shows a small sedate jet within a large black basin. Make the constant slightly larger than $1/e$, and the picture changes dramatically. The Julia set explodes from a relatively small piece of the plane into two spiraling galaxies that fill the plane. Similarly dramatic changes occur when sin z is multiplied by various values of a constant ranging from $1 + 0.05i$ to $1 + 0.8i$. As the imaginary part of the constant grows, the basin of attraction disappears (*see Figure 6.14*).

Computing what happens to orbits across the whole complex plane for various values of the parameter k can get quite tedious. For transcendental functions, mathematicians have found a shortcut for telling when and where a given system bursts into chaos. In the case of systems such as ke^z or k sin z, only certain critical orbits need be followed. For the exponential function, this is the orbit of 0; for the sine function, these are the orbits starting at $\pm\pi/2$. Those points correspond to the location of maxima and minima in the wavelike sine function. Although the sine function has infinitely many critical points, there are only two different values. If all critical orbits go off toward infinity, the dynamical system is completely chaotic on the whole plane. That's easy to check for ke^z. The starting value zero tends to infinity only if k is greater than $1/e$.

All of Devaney's results were suggested initially by computer experimentation, and many features of the behavior of transcendental functions were later established using more traditional mathematical techniques. Further explorations have located bursts in other families of complex functions. For example, a burst occurs in the ik cos z family when k changes from 0.67 to a slightly larger value. Another strikes when k shifts from 0 to a larger number in the family $(1 + ki)$ sin z. Rigorous mathematical proofs verify the occurrence of each burst.

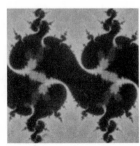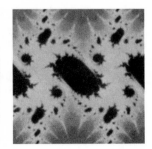

Figure 6.14 As c grows from $1 + 0.05i$ to $1 + 0.8i$, the function c sin z bursts into chaos. The dark areas represent stable regions.

Each of the many complex analytical functions known seems to have its own characteristic behavior. Just at the beginning of their studies of iterated complex functions, mathematicians can look forward to carrying their explorations across whole classes of functions, searching for common elements in their behavior. They also see value in extending the study of iterated functions into higher dimensions.

— — — — — — — — —

The Mandelbrot set and all the other curious objects that arise in mathematical investigations are much more than playthings. They offer ways of exploring the behavior of dynamical systems, in which equations express how some quantity changes over time. Such equations arise in calculations of the orbit of a satellite, the flow of heat in a liquid, and countless other situations. Indeed, dynamical systems are the way scientists model the world, and they are of central importance in both mathematics and science.

Chaotic behavior is a manifestation of a sensitive dependence on initial conditions. Many equations display such a sensitivity, as do a variety of physical phenomena. We can find evidence of this intrinsic unpredictability in attempting to forecast the weather, determine the motion of a passenger on an amusement park ride such as the Tilt-A-Whirl, and even calculate the positions of asteroids and planets far into the future.

At the same time, this sensitivity affords a remarkable opportunity. It means that just as small disturbances can radically alter a chaotic system's behavior, tiny perturbations can also stabilize such activity. In a 1990 theoretical paper, Edward Ott, Celso Grebogi, and James Yorke of the University of Maryland noted that this extreme sensitivity is often regarded as an annoyance. "Yet it provides us with an extremely useful capability without a counterpart in nonchaotic systems," they argued. "In particular, the future state of a chaotic system can be substantially altered by a tiny perturbation. If we can accurately sense the state of the system and intelligently perturb it, this presents us with the possibility of rapidly directing the system to a desired state."

Researchers have already succeeded in applying such an approach to subdue the fluctuating output of high-power lasers and the erratic activity of electrical circuits. The most intriguing possibilities for control may involve biological systems, whether to stabilize irregular heartbeats or to turn on and off seemingly random patterns in the electrical activity of neurons.

7

Life
Stories

Imagine an immense checkerboard grid stretching as far as the eye can see. Most of the checkerboard's squares, or cells, are empty; a few are occupied by strange beings—creatures very sensitive to their immediate neighbors. Their individual fates teeter on numbers. Too many neighbors means death by overcrowding and too few death by loneliness. A cozy trio of neighbors leads to a birth and a pair of neighbors to comfortable stability.

At each time step, this cellular universe shuffles itself. Births and deaths change old patterns into new arrangements. The patterns evolve—sometimes into a static array that simply marks time, sometimes into a sequence of shapes repeated again and again, sometimes into a chain of arrangements that propagates throughout the checkerboard universe.

——————————— The Game of Life ———————————

The mathematical game called "Life" generates a remarkably diverse array of thought-provoking patterns and scenarios. Invented in 1970 by the British mathematician John H. Conway, now at Princeton University, it vividly demonstrates how a set of simple rules can lead to a complex world displaying a rich assortment of interesting behavior. Conway's aim was to create a cellular pastime based on the simplest possible set of rules that would still make the game unpredictable. Moreover, he wanted the rules to be complete enough so that once started, the game could play itself. Growth and change would occur in jumps, one step inexorably leading to the next. The result would be a little universe founded on logic, in which everything would be predestined, but there would be no obvious way for a spectator or player to determine the fate of future generations except by letting the game play itself out.

To find appropriate rules, Conway and his students at Cambridge University in England investigated hundreds of possibilities. They did thousands of calculations, looking at innumerable special cases to expose hidden patterns and underlying structures. They tried triangular, square, and hexagonal lattices, scribbling across acres of paper. They used large numbers of poker chips, coins, shells, and stones in their search for a viable balance between life and death.

The game that they came up with is played on an infinite grid of square cells. Each cell is surrounded by eight neighbors, four along its sides and four at its corners. It is initially marked as either occupied or vacant, creating some sort of arbitrary starting configuration. Changes occur in jumps, with each cell responding according to the rules. Any cell having two occupied cells as neighbors stays in its original state. A cell that is alive stays alive, and one that is empty stays empty. Three living neighbors adjacent to an empty cell leads to tricellular mating. A birth takes place, filling the empty cell. In such a neighborhood, a cell already alive continues to live. However, an occupied cell surrounded by four or more living cells is emptied. Unhappily, death also occurs if none or only one of an isolated living cell's neighbors is alive (*see Figure 7.1*). These simple rules engender a surprisingly complex world that displays a wide assortment of interesting events and patterns—a microcosm that captures elements of life, birth, growth, evolution, and death.

The game was first introduced to the public in October 1970, in Martin Gardner's "Mathematical Games" column in *Scientific American*. It aroused tremendous interest, and the game became an addictive passion for many people. Because it was relatively easy to implement as a computer program, it also quickly became a favorite computer exercise. All kinds of people—students and professors, amateurs and professionals—spent years of computer time following the evolution of countless starting patterns.

"Life" aficionados gleefully pursued elusive arrangements and searched for unusual types of behavior. Many different forms evolved on the checkerboard and were painstakingly cataloged, sporting evocative names such as pipe, horse, snake, honeycomb cell, ship, loaf, frog, danger signal, glider, beacon, powder keg, spaceship, toad, pinwheel, and gun. Some of these arrangements vegetated in a single contented state, and others pulsated, switching back and forth between one configuration and another (*see Figure 7.2*). The possibilities were endless, and the game presented a variety of intriguing mathematical puzzles. For example, are there patterns that can have no predecessor? Several such "Garden of Eden" arrays were eventually discovered (*see Figure 7.3*).

Other investigations revealed that while a given pattern leads to only one sequel pattern, it can have several possible predecessors (*see Figure 7.4*). Thus, a particular configuration can have a number of different

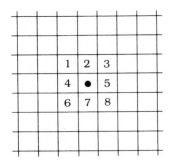

Figure **7.1** A cell's eight nearest neighbors have a strong influence on its behavior.

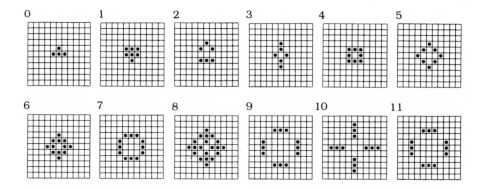

Figure 7.2 One simple pattern evolves over time into a sequence that alternates between two different forms.

pasts but only one future. That makes it difficult for a viewer, glued to a computer screen, to backtrack if a particularly interesting pattern appears fleetingly during the course of a run. There is no guaranteed way to travel backward in time to recreate a past "Life."

The computer also brought animation to the game. A rapidly computed sequence of generations could be viewed as pulsating shapes, creeping growths, lingering dusts, fragmenting forms, and chaotically dancing figures. More recently, enthusiasts have adapted Conway's game for various surfaces other than an infinite plane. Players can now follow the game on the surface of a cylinder, a torus, or even a Möbius strip. They can also pursue the creatures of "Life" on structures in three and higher dimensions.

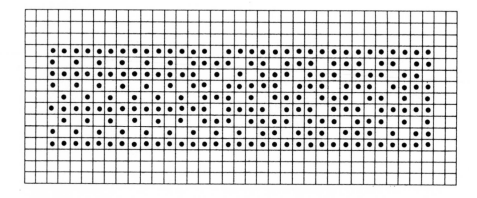

Figure 7.3 A Garden of Eden array.

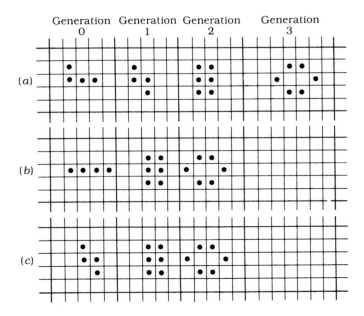

Figure 7.4 Different starting states can lead to identical vegetating states.

The attraction of Conway's original game, and the chief reason for its popularity, is that although it is completely predictable on a cell-by-cell basis, the large-scale evolution of patterns defies intuition. Will a pattern grow without limit? Will it settle into a single stable object? Will it send off a shipload of colonists? Conway managed to balance the system's competing tendencies for growth and death so precariously that "Life" is always full of surprises.

Going with the Flow

Conway's game of "Life" is an example of a cellular automaton, the mathematical equivalent of an array of simple robots programmed to do only certain tasks. The mathematician starts by setting up a field of cells, typically in a checkerboard or honeycomb pattern. Each cell is allowed to exist in one of several possible states. A set of rules specifying how neighboring cells influence each other determines how these states change from one moment to the next. The resulting transitions can be visualized on a grid and strung together into a movie.

In general, each cell of these model universes updates itself independently, basing its new state on the appearance of its immediate surroundings and on some shared laws of change. Governed by sets of simple rules, cellular automata can be used to generate animated graphic images, to design computers, and to simulate physical, biological, and even sociological systems. They have become important mechanisms for investigating pattern formation, evolution, and artificial life—forms that exist only in the computer yet mimic certain aspects of the behavior of living organisms.

The concept of a cellular automaton was introduced in the 1950s by the mathematicians John von Neumann (1903–1957) and Stanislaw Ulam (1909–1984), who succeeded in proving that an abstract pattern could create a copy of itself by mechanically following a set of fixed rules. Their success demonstrated that self-reproduction isn't a property only of living things. Self-reproduction is also possible in the context of physics and logic. That result conjures up a surreal image of endless assembly lines of robots building robots designed for building robots.

In von Neumann and Ulam's scheme, each cell represents an automaton, or machine, that can shift into any one of a finite number of states. For example, a two-state machine could be either on or off, and a five-state machine would have five different settings. The choice of machine type would depend on the application. For his self-reproducing automaton, von Neumann needed 29 states. The fate of each of the 200,000 or so cells in his self-reproducing pattern was influenced only by its own history and by the states of its four nearest neighbors.

Interesting as mathematical curiosities and games, cellular automata are beginning to gain importance as simple models of physical phenomena, ranging from the turbulence a protruding stick generates as it interrupts the smooth flow of water to the spread of an epidemic when a contagious disease claims one victim after another. The idea is to find a way to describe a physical process compactly so that the system's main features can be followed and its future behavior predicted without having to duplicate the system in all its detail.

Traditionally, theorists have used explicit mathematical equations to represent physical processes. Physical laws are often expressed in the form of differential equations, which specify how variables change over time. Solving the equations produces formulas that can be used to predict a system's behavior. For example, Newton's laws of motion provide good estimates of where the planet Jupiter will be several years from now or how long a dropped coin takes to reach the ground.

However, many equations, such as those describing fluid flow are difficult or even impossible to solve exactly. To surmount this obstacle, applied mathematicians over the centuries have developed various procedures and algorithms for finding approximate solutions. The rapidly growing use of computers, which rely on approximation methods to solve equations, has focused even more attention on these methods. Recent improvements have led to efficient computer programs for solving such engineering problems as designing aircraft. Approximation methods have also led to puzzling new concepts, such as chaos, as discussed in Chapter 6.

The mathematician Stephen Wolfram, now known for developing software for performing highly sophisticated mathematical operations, was one of the pioneers in the use of cellular automata to represent physical phenomena. Wolfram argued that his scheme is better suited to modern digital computers than the various methods normally used to handle models based on differential equations.

In a sense, Wolfram modified the way a theory is constructed so that he could take the best possible advantage of an available tool. That's like changing a recipe to suit the kitchen implements and ingredients on hand. The advent of massively parallel processing computers, in which networks of simple processors allow computations to be divided up so that each processor performs the same operations on different pieces of data at the same time, makes the idea even more attractive.

In 1984, Wolfram explained his approach in an article in the journal *Nature*:

> Cellular automata are examples of mathematical systems
> constructed from many identical components, each simple, but
> together capable of complex behavior. From their analysis, one
> may, on the one hand develop specific models for particular
> systems, and, on the other hand, hope to abstract general
> principles applicable to a wide variety of complex systems.

Wolfram's idea, like that of von Neumann and Ulam, was to start with a lattice of sites (like the squares of a checkerboard or the hexagonal cells in a honeycomb), with each site carrying a discrete value chosen from a small set of possibilities. At every time step, these cell values are updated according to rules that depend on the values of neighboring sites.

The simplest types of cellular automata are narrowly confined to one dimension. A typical computer experiment starts with a line of, say, 600 sites. Each site has a value of either 1 or 0. Placing an initial pattern on the line, computing successive generations, and displaying them as black (1) or white (0) squares in sequence on a computer screen show how the pattern evolves. Like an unfurling banner, the pattern's successive rows, marking each new generation, march down the computer screen.

What happens depends on the particular rules employed. For example, the next state of a given cell may be determined by the sum of the states in the cell's neighborhood. By one such rule,

Neighborhood sum	5	4	3	2	1	0
Cell's next state	0	1	0	1	0	1

In this case, a neighborhood consists of a cell and two sites to the right and two sites to the left of the given cell. If each of the five cells has a value of 1, the total is 5. According to the table, the central cell's value in the next generation will be 0. The same procedure is applied to each cell in the string.

Altogether, with two states and five-site neighborhoods, there are 64 different ways to assign values to cells in succeeding generations. Wolfram explored all these possibilities and many others. His empirical study suggests that patterns evolving from a simple "seed," consisting initially of just a few nonzero sites, fall roughly into four categories (*see Figure 7.5*). Some rules don't lead to anything—the starting patterns simply die out. For other rules, the seed patterns evolve to a fixed, finite size. A third group of rules generates patterns that keep growing at a fixed rate. These patterns are often self-similar, and many have a fractal dimension of 1.59. Finally, some rules generate patterns that grow and contract quite irregularly and unpredictably.

Similar studies can be done on patterns initially consisting of, say, 100 cells, each one randomly assigned a value of 1 or 0. Again, the patterns produced by various rules tend to fall into four classes, reminiscent of the mathematical behavior of nonlinear differential equations or iterated functions under various conditions (*see Figure 7.6 and Color Plate 14*).

About one-quarter of the rules that Wolfram studied leads to patterns that converge after a finite number of generations into a single, homogeneous state that endlessly repeats itself. Several of these look like vertically striped hallway carpets. Close to 16 percent of the rules lead to a number of either unvarying or simply repeating patterns. More than half of the rules generate patterns that never develop any structure. The patterns look chaotic. A small number of rules generates patterns that develop complex, slowly changing subpatterns, some of which are remarkably persistent.

Wolfram also discovered a novel, efficient way of generating random numbers. Some rules produce patterns that look completely random. In at least one case, because it satisfies several important mathematical tests of randomness, the central column from such a pattern can be used to specify

Figure 7.5 Three of four classes of patterns generated by the evolution of one-dimensional cellular automata from simple "seeds."

Figure 7.6 Patterns generated by four different rules. In each pair of patterns, the upper one is obtained starting with a single nonzero site; the lower indicates what happens when a set of randomly chosen initial values is used.

(continued)

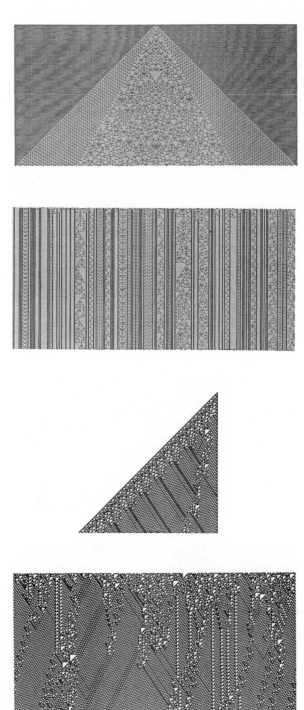

Figure 7.6 *(continued)*

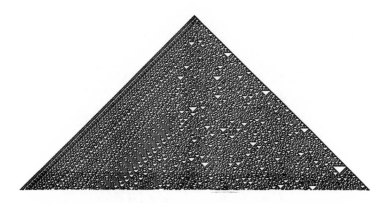

Figure 7.7 A cellular automaton generates an apparently random string of 1s and 0s in the center column.

a string of random 1s and 0s (*see Figure 7.7*). In a sense, this particular cellular automaton hides, or "encrypts," the original data. Given only the output sequence after many generations, it would be very difficult to deduce the original seed.

A cellular automaton that appears to generate random numbers mimics the kind of behavior shown by many mathematical quantities. For example, it's easy to write down an equation that specifies the value of π, the ratio of a circle's circumference to its diameter; that equation can be used to compute π to as many digits as desired. But once the computation is done, the sequence of digits appears random, for all practical purposes. Tests on the 52 billion or so known digits of π have so far failed to turn up any evidence of an underlying pattern in the digits. Given only the digits of π, there would be no way to reconstruct the equation that defined and generated the number in the first place.

When using a cellular automaton as a model of a physical system, the challenge is to capture the key features of the phenomenon in as simple a model as possible. Wolfram built his cellular-automaton models mainly by trial and error, testing various grids and sets of rules until he found a cellular automaton that produced the kind of large-scale behavior he wanted to see.

In Wolfram's model of a flowing liquid, particles move in discrete steps along the links of a hexagonal lattice, bouncing from cell to cell. Each link supports one particle. Particles collide and scatter according to rules that ensure that the total number of particles doesn't change and that the total momentum carried by the particles is conserved. To satisfy the second constraint, Wolfram keeps the total number of particles roughly the same in each direction on the lattice (*see Figure 7.8*).

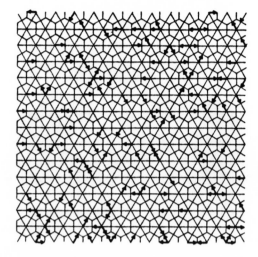

Figure 7.8 In this cellular-automaton fluid model, each arrow represents a discrete particle on a link of the hexagonal grid.

Computer simulations of 10 million particles bouncing around in a grid show what happens. When the particle motions are averaged over large regions, large-scale flow patterns appear. For example, a cylinder (represented in the two-dimensional grid as a circle) moving through a fluid at rest sheds the characteristic, miniature whirlpools seen in corresponding water-tank experiments (*see Figure 7.9*).

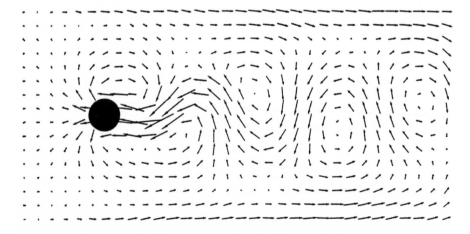

Figure 7.9 A fluid-flow pattern is obtained from a simple two-dimensional cellular automaton.

A similar approach can be used to model other physical processes, such as diffusion. In this case, particles bunch in a regular array at the center of a grid. Appropriate rules allow the particles to spread out into an apparently random arrangement. Wolfram's computer simulations showed that on the average the number of particles in a given region after a certain time interval matches the predictions of equations commonly used to represent diffusion.

Another striking application of cellular automata involves models of the kind of branching crystal growth that leads to snowflakes and other dendritic crystal forms. Governed by an appropriate two-dimensional rule, which accounts for the inhibition of growth at newly formed sites, a small seed can grow into a fractal form that resembles a snowflake.

One simple way to grow a snowflake on a computer, developed by Norm Packard, is to start with a small ice hexagon at the center of a hexagonal grid. A blank hexagon turns to ice if there is an odd number of ice hexagons surrounding it. No ice hexagon forms if the cell has an even number of ice hexagons as its nearest neighbors. More complicated rules lead to more intricate snowflake patterns (*see Color Plate* 15).

The advantage of cellular-automata models is their simplicity. They mimic natural systems whose basic parts are each very simple but whose overall behavior can be very complex. What still isn't clear is how useful these models are for predicting the large-scale, long-term behavior of physical systems. There's even a possibility that it would take so long to compute an answer in order to make a prediction that it would be easier to watch the system itself to see what it does. Prediction becomes impossible. As Wolfram noted:

> Much has been discovered about the nature of the components
> in physical and biological systems; little is known about the
> mechanisms by which these components act together to give
> the overall complexity observed. What is now needed is a general
> mathematical theory to describe the nature and generation of
> complexity.

Forest Fires, Barnacles, and Trickling Oil

Introducing an element of chance into the rules governing cellular automata brings a new dimension into studies of complex systems. Such mathematical models appear to mimic a wide range of rather unpleasant occurrences, from the spread of disease and forest fires to an invasion of crabgrass. These interacting particle systems, as mathematicians prefer to call them, simulate the irregular spread of infestations as they jump from one victim to the next. Depending on the specific rules, all kinds of scenarios are possible.

The study of interacting particle systems is a relatively new activity for mathematicians. It began in the late 1960s as a branch of probability theory. Since then, this area has grown and developed rapidly, establishing unexpected connections with a number of other fields. The first examples of interacting particle systems were suggested by research in statistical mechanics. Physicists wanted to understand how a collection of wandering or randomly scattered particles can suddenly organize itself, as happens when a liquid solidifies or a material is magnetized.

The mathematical models developed to simulate such phase transitions proved to be a rich source of inspiration. The idea was to look at what happens to particles scattered across a grid when each particle is allowed to interact with its neighbors, according to certain rules. It became clear that models with a very similar mathematical structure could also be useful for studying neural networks, tumor growth, ecological change, and the spread of infections. Moreover, the dynamical behavior of such models suggested intriguing mathematical questions. Mathematicians became interested in how certain models evolve over time and began searching for unusual types of behavior.

Rick Durrett, one the leaders in this new area of mathematics, says, "Mostly, it's a mathematician's game of seeing what happens." The mathematician defines the rules, sets up the game board, and lets the game play itself out.

One example of an interacting particle system is a model that resembles the inexorable spread of a raging forest fire, as it fans out in a blazing ring, consuming fresh timber at its outskirts and leaving a burnt residue within. The rules for a mathematician's forest fire are simple. The playing field is a checkerboard grid on which each cell represents a tree. The fire begins as a single marked cell or a small cluster of cells at the grid's center. At each time step, a burning cell has a certain probability of spreading the fire to its four nearest neighbors, unless those neighbors have already been burnt.

The same rules also lead to a rough model of the spread of an infectious disease such as measles. In this case, each cell represents an individual who is either healthy, ill, or immune at any given moment.

The simplest set of rules mathematically worth investigating makes the unrealistic assumption that the fire (or infection) in a given cell lasts for only one unit of time. At each step, the toss of a coin or a similar randomizing procedure decides whether a certain burning cell spreads its flames to each of its neighbors. Hence, at any given time, a computer screen would show three types of cells: those that are burning, those already burnt up, and those that have so far escaped unscathed. The burning cells usually form an irregular broken ring that gradually expands as time goes on.

Mathematicians, particularly probabilists, are interested in how the process depends on the probability of transmission from one cell to another. They find that when a cell has only four chances in 10 (a probability of .4) of passing on the fire to a given neighbor, the fire eventually dies out. If it has six chances in 10 (a probability of .6) of doing the deed, the fire continues to spread. The critical probability that establishes the dividing line between these two types of behavior turns out to be .5.

The shape of the ring also appears to depend on the transmission probability. At a probability of .5, the ring's outer edge looks like a fractal, having an incredibly convoluted boundary that when magnified, instead of looking smoother, appears to be equally complicated on every scale. Theorists have no idea why a fractal should appear at this critical value. By the time the probability reaches .6, the ring is smoothed out into a circle. At even higher probabilities, the ring takes on the shape of a square, reflecting the underlying grid (see Figure 7.10).

A similar model can be applied to the spread of measles. Before the development of a vaccine for the measles virus, the disease would sweep through a community of school-aged children much like a forest fire. Just as a spark from a flaming tree might ignite its neighbor, physical contact with virus-laden mucus passed the disease on from one child to another.

Working with Simon Levin, a mathematical ecologist at Princeton University, Durrett used an interacting particle system model to generate animated sequences showing the spread of measles. In their model, an individual could be susceptible to being infected, actually infected with the disease, or immune to further infection for a certain period. Susceptible persons became infected at a rate proportional to the number of infected neighbors. The trick is getting the rate right to match epidemiological data. If the transmission rate is high, the entire population dies out. A very low rate means that the disease disappears quickly.

Durrett's model shows oscillations reminiscent of the recurrence at 20-year intervals of various flu viruses. Such oscillations occur when the rate at which an immune individual becomes susceptible again is low. The results agree with data suggesting that a population of 250,000 is needed to keep measles from dying out.

Techniques involving interacting particle systems and related cellular automata have been used to explore the spread of impulses across the muscle tissue of the heart, to study the characteristics of zones where the territories of two related species of plants or animals overlap, and to explain the empirically observed relationship that the number of species present in a given area is proportional to the area raised to a certain power. The species-area power law relationship is important for understanding how natural communities are organized or how the size of a reserve affects biodiversity.

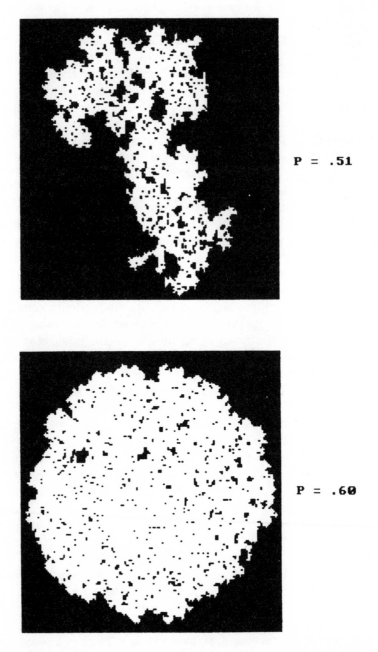

P = .51

P = .60

Figure 7.10 The growth pattern for the forest-fire model looks like a fractal when the probability of transmission is near .5 but is roughly circular at a higher probability.

Oriented percolation, another example of an interacting particle system, resembles the downward trickle of a surface pool of oil through an underlying bed of sand or soil. Fingers of oil penetrate this layer if the ground is porous enough. The trick is to find the critical probability at which a sufficient number of air spaces are present so that an open pathway exists. Because no one knows precisely what this critical probability is, theorists must resort to computer simulations to get a feel for what the process looks like.

The oriented percolation model can be pictured as a gigantic diamond-shaped network of connected pipes (*see Figure* 7.11). A valve in each section of pipe in the grid may be either open or closed. At the top of the network is a reservoir of fluid. If all valves are open, fluid will flow down through the network of pipes. If all valves are closed, no flow occurs.

What if some valves are open and others closed? Clearly, no fluid will flow until enough valves are open. By studying the results when the probability of a particular valve being open lies somewhere between 1 (all valves open) and 0 (all valves closed), mathematicians can determine the critical probability at which flow is first established.

Computer experiments show that when the open-valve probability is .55, the process dies out. This means that on the average, when 55 out of the 100 valves are open, fluid gets into the pipes but doesn't get very far before all paths for fluid flow are blocked. As the probability that a valve is open increases, the fluid penetrates deeper. Percolation—long-distance flow—is established when the probability of an open valve gets close to .645 (*see Figure* 7.12). However, a mathematical proof that this number must be the critical value has been elusive.

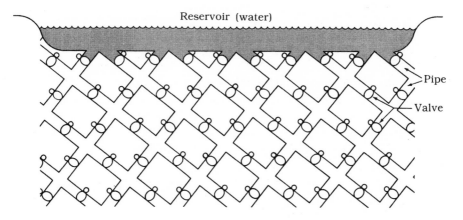

Figure 7.11 How far water penetrates into this network of pipes depends on which valves are open.

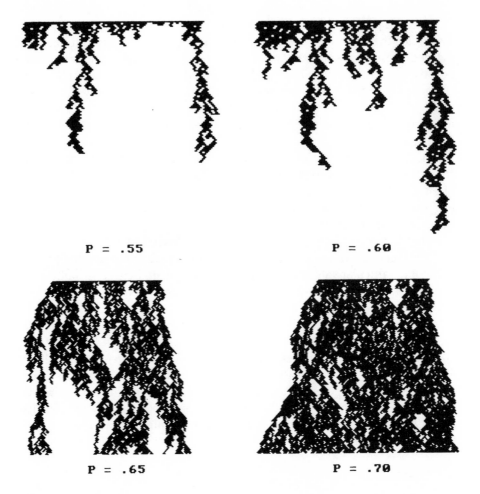

P = .55   P = .60

P = .65   P = .70

Figure 7.12 Changing the probability that a valve is open changes how deeply a fluid penetrates the matrix.

A similar, two-dimensional model can be set up for the spread of a plant species or the propagation of a population of immobile animals, such as barnacles or mussels. This particular model is named for the British mathematician Daniel Richardson, who first suggested it in a 1973 paper. Partly because there is no provision for death, the model's pattern always grows to cover the entire plane, leaving only a few holes near its fringes. Theorists are interested in the times at which a growing shape exceeds a certain boundary and in how this time depends on the probability of an occupied cell sending an offspring (or root) to an adjacent, unoccupied space (*see Figure* 7.13).

The examples illustrate just a few of the infinite number of possibilities that could be investigated. Even in a problem as simple as the forest fire

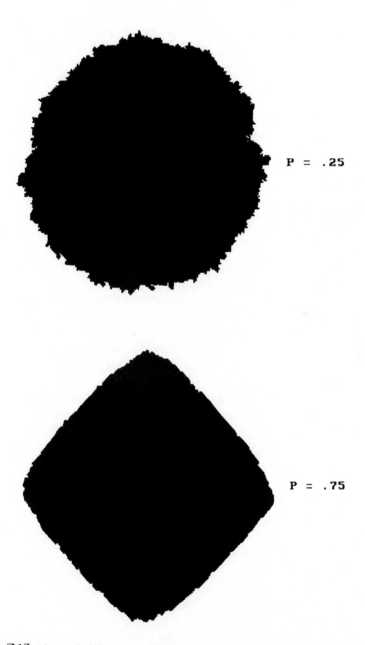

P = .25

P = .75

Figure 7.13 The probability of sending offspring to a neighboring cell affects the geometric shape of the pattern.

model, the number of possibilities is striking—the rules can be changed to involve a different number of neighbors; the burning stage could last for more than one unit of time; the grid itself could be hexagonal or triangular instead of square. The possibilities appear limited only by the researcher's imagination.

How do mathematicians decide which models are worth studying? Some models are suggested by studies in other fields, such as physics and biology. Others show some type of novel behavior. In general, most models do nothing interesting. Only a few do something special, and these are the ones that mathematicians seek out and study. And there's another barrier. Many candidate models are too complicated to be analyzed mathematically. Mathematicians keep trying rules until they find a system they can conveniently handle.

Both the emphasis on diversity and the difficulty that mathematicians have in rigorously proving general statements about these models lead to what seems a curiously fragmented type of mathematics. The unity of the subject doesn't come from general theorems, as in other fields of mathematics. Instead, it comes from the same techniques being used to analyze a variety of models.

The way various interacting particle systems behave in different dimensions is a topic that has produced some interesting results. Such models need not be restricted to planar cells or even to three-dimensional blocks. Although difficult to visualize, these models can be studied in any dimension. Mathematicians have found out that the behavior of models changes in different dimensions. Sometimes, they find a critical dimension, which means that in lower dimensions one thing happens and in higher dimensions something else happens.

In fact, the behavior of most models seems to get simpler when the dimension gets larger. Because higher dimensions have so much extra room, the interactions between particles get weaker. Particles tend to be spread more widely. Frequently, higher-dimensional versions of various processes behave almost as if there were no interactions. They begin to look a lot like processes such as random walks, traditionally studied in probability theory and made up of independent steps.

With the increasing use and greater sophistication of computer simulations, it's possible that interacting particle systems may eventually find uses in fields such as ecology and geology. Durrett remarked:

> Some may criticize these models for their simple-minded approach
> to modeling the interactions of individuals and, indeed because
> of this, they cannot be used to obtain quantitative conclusions.
> However, there is a growing list of examples to show that these
> models can provide important insights into the rhythms and
> patterns of nature.

—————— — — - Travels of an Ant — —————— ———

It's easy to brush ants aside. Scurrying across a sidewalk, navigating the ridges of a wrinkled picnic blanket, or crowding around a crumb on the kitchen floor, these spindly, communal insects typically attract scant attention to their varied doings—except when they get in our way. Yet in their foraging and social organization, ants display remarkable behavior worthy of detailed study.

James Propp, however, is interested in the activities of a different sort of ant. A mathematician at the Massachusetts Institute of Technology, he has spent the better part of a decade tracking an imaginary critter—a virtual ant—roaming an infinite checkerboard. Exhibited on a computer screen, this mathematical ant blindly follows the dictates of simple rules, tracing a winding path across the plane. Propp doesn't insist on a connection between the actions of the carefully shepherded, simple-minded ants in his simulations and the multifarious antics of ants in the wild. What intrigues him are the intricate patterns and symmetries that can emerge out of a bare-bones mathematical framework.

Like the game of "Life," Propp's ant universe is an example of a cellular automaton, but the rules these virtual ants obey operate a little differently. In this type of cellular automaton, a change of state occurs in only one cell at a time instead of across the board with each step.

Suppose all the cells of a particular ant universe begin in one of two possible states, designated 0 and 1 (or white and black when visualized on a computer screen). Initially, the virtual ant sits on a cell, facing one of the four compass directions. The ant then moves in that direction to an adjacent cell. When it arrives at its new location, the ant is programmed to change its heading by 90 degrees to the left if it lands on a 0 cell or 90 degrees to the right if it lands on a 1 cell. As it leaves, it causes the cell's state to switch from 0 to 1 or from 1 to 0. Thus, on its next visit to this particular cell, the ant will find an altered state and behave accordingly.

Depending on the initial distribution of states, the ant appears to perform an intricately choreographed dance across the plane. This is a cybercritter that can't stick to the straight and narrow. Constantly changing direction, it interacts with its own path—its own history—as it treks from cell to cell.

Propp didn't actually invent the basic ant universe. He originally came across it in the work of Christopher G. Langton of the Santa Fe Institute in New Mexico. In the mid-1980s, Langton created a number of these simulated ant farms, including examples in which several ants move at the same time, to explore how cellular automata might serve as models of various processes characteristic of living systems.

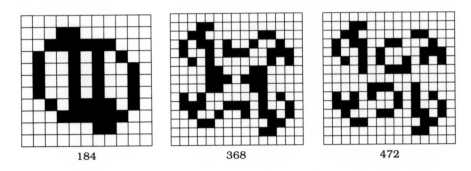

184   368   472

Figure 7.14 The track left by a two-state ant can sometimes appear symmetrical, as seen here at steps 184, 368, and 472.

The special case in which all cells are initially in the 0 state provides a glimpse of how a virtual ant typically behaves. From time to time, the ant returns to its starting point, leaving a symmetrical pattern of cells, all in the 1 state, in its wake (*see Figure* 7.14). At other times, the pattern gets scrambled. This sort of symmetry is not that of an idealized snowflake, which remains completely symmetrical from beginning to end. Rather, it is a recurrent symmetry that is repeatedly destroyed and recreated. Hence, most of the time, the ant's track looks disorganized. At certain intervals, however, a symmetrical pattern emerges, only to be broken up again.

Such behavior intrigued Propp. Picking up where Langton left off, he extended the original model to an infinite array of cells and many more steps. To determine the potentially complicated consequences of these extremely simple rules, the ant had to do its dance for at least 10,000 steps before one could determine its long-term destiny.

Propp was surprised to find that after thousands of steps of rather chaotic, aimless movement, the ant would suddenly appear to make up its mind about where it wanted to go. It would head off, creating a broad track—a highway—in one of the four possible diagonal directions, never to return to its starting point (*see Figure* 7.15). He found the same phenomenon occurs for many initial conditions, although precisely when the highway starts and in which direction it proceeds varies. This startling behavior is somehow encoded in the rules but becomes evident only when the ant is activated for a sufficiently long period.

Meanwhile, Greg Turk, a computer scientist at the University of North Carolina, had independently developed the same kind of simulation while experimenting with a special type of Turing machine (*see Chapter* 8). Such a machine serves as a convenient mathematical model of computation,

Figure 7.15 After about 10,000 steps, a two-state ant suddenly starts building a highway.

performing a sequence of basic operations to accomplish anything that a modern digital computer can do.

Normally, a Turing machine can be pictured as a device that reads symbols—one at a time—from a row of cells on an infinitely long tape. The given rules specify which symbol to write in the current cell, in which direction to move the tape, and what instruction to follow next. Turk extended this model to two dimensions, freeing the read-write device to traverse a square grid. Programmed appropriately, his "tur-mite" could generate a variety of patterns, including spirals, symmetrical shapes, and the highways that Propp had discovered.

Taking such an approach, Serge E. Troubetzkoy of the University of Alabama and Leonid A. Bunimovich of the Georgia Institute of Technology proved what has become the fundamental theorem of cybermyrmecology—

the study of virtual ants. They demonstrated that an ant's track is unbounded. This means that no matter what the initial conditions, the ant never wanders about in such a way that its universe—the pattern of 0s and 1s, as well as the ant's position and orientation—returns precisely to its initial state. In other words, the ant universe never repeats itself.

This theorem applies even when cells in the ant universe can exist in more than two states. For example, a cell may cycle through five states, going from 0 to 4 and back to 0, over the course of successive visits by the ant. The ant itself is programmed to turn left or right according to a rule-string such as LLRRL, with one turning instruction for each of the five possible cell states.

Given enough time, a virtual ant will escape from any finite region. Along the way, depending on the particular rule-string it obeys, the ant often manages to build a sequence of ever larger symmetrical structures before transforming its world into a chaotic jumble or a perpetual highway.

How recurrent symmetry arises was one of the first mathematical mysteries that confronted Propp and others enthralled by this ant universe. With no memory and no ability to plan, how did a single-stepping, short-sighted ant know what to do to maintain the resemblance between the way things look at one location and the way things look far away?

One of those who joined the hunt for a solution to this conundrum was Bernd Rümmler, a mathematician working as a computer programmer at an insurance company in Göttingen, Germany. Limitations of the computer system to which he had access forced him to look for another way of representing an ant's movements on a computer screen. Instead of using blocks of color or shades of gray to represent the state of each cell, he marked the squares with two quarter circles in opposite corners. Arrows showed the direction of travel for either a left turn or a right turn. Such squares are known as Truchet tiles. Positioned next to each other to form a grid, they join together in such a way that the markings create circles and wavy lines across the plane (*see Figure* 7.16 *and Color Plate* 16).

The technique provided an entirely different way of visualizing what a virtual ant does. In essence, the ant follows whatever curve it's on, flipping or not flipping a tile as required by its program when it exits a cell to travel to the next one. From the patterns of the curves winding from cell to cell, it was easier than before to tell where the ant had been—and especially where it was going. The curves enabled Rümmler and others to see that, between the moments when a symmetrical structure appears in the ant's universe, the symmetry is not completely destroyed; rather, it goes "underground," only to reappear a moment later in a larger

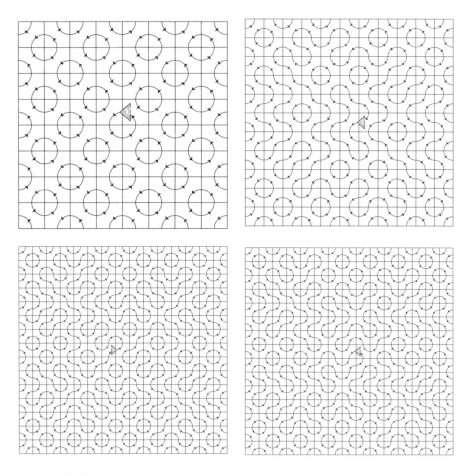

Figure 7.16 Initial configuration of the universe of a two-state ant (*shaded triangle*), as represented using Truchet tiles (*top left*). After 184 (*top right*), 368 (*bottom left*), and 472 (*bottom right*) steps, the ant has reconfigured the tiles to create lengthy connected curves, or contours.

pattern. In other words, the recurrent emergence of symmetry is embedded in the curved tracks of a curiously switched railroad that the ant simply follows.

Many mysteries of the ant universe remain unsolved. Lots of variations of the basic rules and the distribution of initial states still haven't been explored. No one has looked seriously at virtual ants traversing a three-dimensional lattice of cubes. Propp himself has pondered ant behavior on a field of hexagons. It's all part of the wonders of ants on the move.

The Gods of Sugarscape

In the ritualized warfare of the game of chess, the board is a miniature battlefield on which opposing commanders in chief marshal their forces. Each playing piece has a particular pattern of allowed movements, and the game's rules shape the battle. The combatants can try out different strategies, directing bold attacks, mounting stubborn defenses, or waging wars of attrition across the grid.

At the Brookings Institution in Washington, D. C., Joshua M. Epstein, a social scientist, and Robert Axtell, a computer modeler, have a playing field of their own on which to audition their ideas. They play a game on a 50-by-50 square lattice on a computer screen, and the playing pieces, or agents, are colored dots occupying some fraction of the squares. Indeed, Epstein and Axtell are actually more like gods than commanders. They define the landscape, set the rules, and characterize the agents. Instead of participating in the action, they step back to observe what happens as the swarming agents, left on their own in these simulations, move about, gather sustenance, reproduce, and die off according to their programmed predilections. From the patterns that emerge, the researchers can glean insights into human social and economic behavior.

"We grow social structures—artificial societies—in the computer," Epstein explains. "We can examine population growth and migration, famine, epidemics, economic development, trade, conflict, and other social issues."

The cyberworld in which Epstein and Axtell's agents dwell is known as Sugarscape. It's a two-dimensional landscape, represented as a square grid, containing two regions rich in a renewable resource arbitrarily called sugar. Every agent is born into this world with a metabolism demanding sugar, and each has a number of attributes, such as visual range for food detection, that vary across the population. The agents move from square to square according to a simple rule: Look around as far as your vision permits, find the unoccupied spot with the most sugar, go there, and eat the sugar. As it is consumed, the sugar grows back at a predetermined rate. An agent's range is set by how far it can see. Every time an agent moves, it burns an amount of sugar determined by its given metabolic rate. Agents die when they fail to gather enough sugar to fuel their activities.

With hundreds of agents roaming the landscape, interesting things begin to happen. Initially distributed at random across the landscape, the agents quickly gravitate toward the two sugar mountains. A few individuals end up accumulating large stocks of sugar, building up a great deal of personal wealth. These happen to be agents that have superior vision and a low metabolic rate and have lived a long time. A few others, combining short vision with a low metabolic rate, manage to subsist at the fringes, gathering just enough to survive in the sugar badlands but not looking far enough to see the much larger sugar stocks available just beyond the horizon (*see* Figure 7.17).

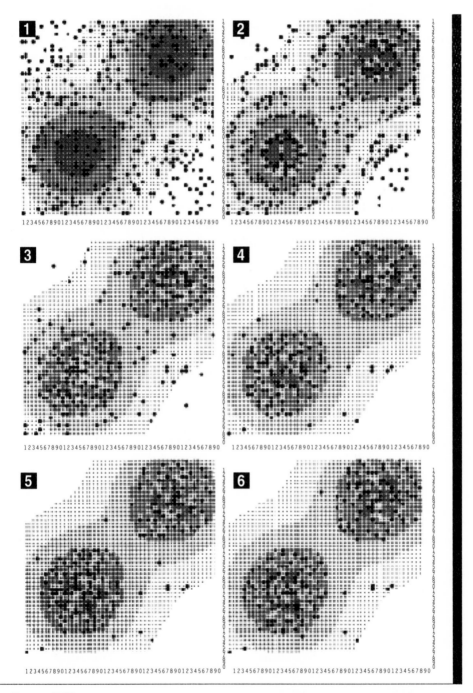

Figure 7.17 This sequence (1–6) begins with agents (*black dots*) distributed at random across a landscape that features two concentrations of a resource arbitrarily called sugar (*gray dots*). As time passes, the resource-seeking behavior of the agents tends to lead them to the sugar mountains. Only a few manage to survive on the fringes.

At its simplest level, the Sugarscape model represents a kind of hunter-gatherer society. But even this rudimentary model reproduces the kind of strongly skewed distribution of wealth generally observed in human societies—where a few individuals hold most of the wealth and the bulk of the population lives in relative poverty.

Introducing sex and reproduction to Sugarscape is as simple as adding to the agent's string of numbers a few bits representing gender. A rule specifies the allowed behavior—an agent must select a neighboring agent at random. If the neighbor is of the opposite sex and of reproductive age, and if one of the agents has an empty neighboring site (to hold offspring), a child is born. The child inherits a mixture of its parents' genetic attributes.

This new dimension enables the researchers to investigate the effect of cultural forces on biological evolution and vice versa. For example, in the absence of any cultural factor, agents with relatively low metabolism and high vision enjoy a selective advantage in Sugarscape. Now, suppose that when an agent dies, it can pass on its accumulated holdings of sugar to its offspring. How does this cultural convention influence evolution? The Sugarscape model suggests that agents who might otherwise have been "weeded out" are given an extra advantage through inheritance. The average vision of the population doesn't increase to the same high level eventually reached in a population where no wealth is passed on.

The researchers can modify their model to observe the emergence of tribes (identified by numerical tags) and the process of assimilation (changing affiliation to join the local majority). Inevitably, there arises a primitive kind of combat, in which agents of two different tribes may plunder each other for sugar. Various combat rules lead to patterns of movement that differ from those produced by the standard "eat all you can find" rule. For example, some combat rules lead quickly to strictly segregated colonies, each clinging to its own sugar peak. In other cases, one side wipes the other out.

The Sugarscape model also offers insights into other phenomena, such as the introduction of trade. In this case, the landscape contains heaps of two resources—sugar and spice. The agents are programmed with different metabolic rates, or preferences, for each of the two commodities. They die if either their sugar or their spice store falls to zero. A mathematical formula called a welfare function allows each agent to compute how close it is to sugar or spice starvation. The agent then strives to gather more of the commodity it needs. An additional system of rules specifies how agents bargain for and exchange sugar and spice according to their needs. These rules enable the researchers to document how much trade transpires and at what price exchanges occur.

When agents are allowed to live forever, as long as they never run out of food, the sugar-spice model shows that the average trade price converges to a stable level. Economic equilibrium emerges, just as textbook economics predicts. However, when Epstein and Axtell make the agents "more human" by giving them finite lives and permitting their preferences to evolve, the price no longer stabilizes and the market never reaches equilibrium. "The assumption that we can let markets produce efficient allocations of capital or resources on their own is deeply challenged by our work," Epstein claims. "We see how brittle traditional economic theory really is."

Epstein and Axtell's Sugarscape simulation is just one example of a wide variety of computer models now being developed on the basis of interactions between agents governed by given rules rather than on equations defining global behavior. The idea is to model from the bottom up—seeing behavior emerge out of interactions among individuals—instead of from the top down—deriving the behavior of individuals from overarching laws. Researchers at the Santa Fe Institute, Los Alamos National Laboratory, and elsewhere have worked out agent-based models of urban transportation systems, insect colonies, business organizations, financial markets, and other situations. Such approaches are also useful in studies of artificial life.

What distinguishes the Sugarscape project is its emphasis on seeing what sorts of socially relevant behavior can emerge from the collective interaction of individuals following the simplest possible rules. "The surprise is that we can grow complex, recognizable behavior with incredibly simple rules and simple agents," Epstein says. Such agent-based modeling shows that social norms can arise out of very primitive behavior, although it doesn't necessarily demonstrate how the norms actually came about.

Epstein insists that although these bare-bones models can't really be used to make specific predictions, they can suggest explanations of some widely observed macroscopic phenomena, from distribution of wealth in typical societies to erratic price fluctuations in markets. "We think of our model as a laboratory for social science," Epstein says. Researchers from a wide variety of disciplines, including economics, biology, demographics, and environmental studies, can use this approach as a research tool to tackle oft-neglected, cross-disciplinary issues like the effects of inheritance on the genetic evolution of a system.

Like others working with agent-based models, Epstein and Axtell must interpret the patterns they observe on the computer screen. After all, their agents are no more than strings of digits and the observed behaviors no more than patterns in a computer's memory. Indeed, it is the act of interpretation that allows such electronic worlds to make contact with their real-world counterparts. In the case of Sugarscape, the model is more a metaphor than a realistic depiction of society. No one literally spends a working day accumulating sugar. The landscape and agent characteristics are simple stand-ins for the more complicated things that occur in the real world.

In one application of their method, Epstein and Axtell have worked with the Santa Fe archaeologist George J. Gumerman and his colleagues to "grow" the Anasazi society, a Native American culture that flourished in the American Southwest for hundreds of years, then suddenly disappeared. The archaeologists have data on weather patterns, crop yields, and other environmental conditions during that period, along with information about the Anasazi culture. The researchers hope that agent-based simulations may shed light on whether environmental or cultural factors were primarily responsible for the society's abrupt decline.

Epstein and Axtell have also collaborated with an economist to study how caste systems, in which a small elite demands more than its fair share, arise in societies. The idea is to get a sense of whether equity comes about naturally in social systems.

The Sugarscape laboratory remains very much in the development stage. Researchers are just starting to examine ways of tailoring this approach to address specified needs and issues in the social sciences, economics, and elsewhere. One key issue in the Sugarscape approach involves how to tweak the model to obtain such phenomena as the emergence of governments. There may be some sort of threshold beyond which researchers can't take a step up in using what are essentially cellular-automaton models to understand human organization without making the agents smarter.

Sugarscape is already an immensely attractive playing field because the limited repertoire of the agents makes it easy to understand, measure, and depict what's going on and why the agents behave as they do. Moreover, typical Sugarscape experiments take only a few minutes on an ordinary desktop computer. By providing insights into population growth, resource use, migration, economic development, conflict, and other global social processes, games played on the Sugarscape grid may help shape the policies needed to direct the future course of society.

———— —— —— - ———— ——

From the game of "Life" to Sugarscape, models based on cellular automata provide insights into how simple interactions between neighboring individuals can lead to complex large-scale behavior. Interpreting the behavior observed in such models, however, raises many of the same issues that come up whenever mathematics is applied. One key question concerns the origin of the rules themselves. In physics, researchers have recourse to fundamental, universal laws, expressed in a mathematical form. In social and biological systems, they can look to history and the principles of evolution to help formulate such rules.

There is a crucial difference between a thing and a mathematical model of the thing. Indeed, the distinction between mathematics and the applications of mathematics often isn't made as clearly as it ought to be. It's quite possible, for example, to calculate correctly the area of a rectangular piece of property just by multiplying the length times the width. One can still get the "wrong answer" in a practical sense, however, because the measurements of the length and width were inaccurate, there was some ambiguity about the boundaries, or if the field wasn't truly rectangular. In mathematical terms, the formula $A = xy$ always works, whether A is area and x and y and length and width or A is force and x and y are mass and acceleration, as in Newton's law of motion. How we interpret the answer, as applied in the real world, depends on the assumptions that went into our selection of the mathematics to use. The same consideration holds for any mathematical model we construct.

8

In Abstract Terrain

Conjecture and proof are the twin pillars of mathematics. Conjecture stands at the cutting edge, in the wild region separating the known from the unknown. Proof stands at the center of mathematics, made manifest as a solid tower built of bricks called mathematical theorems. The task of mathematicians is to develop conjectures—guesses or hypotheses about mathematical behavior—that they can then attempt to prove. On this basis, mathematics grows, stretching into new fields and revealing hitherto unseen connections between well-established domains.

The mathematician William P. Thurston noted: "As mathematics advances, we incorporate it into our thinking. As our thinking becomes more sophisticated, we generate new mathematical concepts and new mathematical structures; the subject matter of mathematics changes to reflect how we think."

Keeping Secrets

Take a generous helping of proof—the essence of serious mathematics—and mix it with a dash of clever computer science. Add a sprinkle of scrambled graphs, throw in a touch of chance, and spice it with the tang of secrecy. The result is a piquant brew that adds up to a playful, foolproof method of keeping a secret. It gives a wary mathematician, caught in the sometimes contentious world of mathematical research, an ingenious way to claim credit for being the first to find a particular proof without having to give away the slightest clue as to what the proof is. All that a rival can find out, until the proof itself is finally unveiled, is that a particular theorem has been proved.

The mathematical basis for such a seemingly impossible scheme is a novel concept called a zero-knowledge proof, first formally defined in 1985. At the root of this development lies the startling notion of a probabilistic, interactive proof. Unlike traditional methods—familiar to any student who has tried to prove one of Euclid's geometric theorems by constructing a chain of statements inexorably leading to the necessary conclusion—the new technique relies on randomness and the interplay between a "prover" and a "verifier" to achieve a practically unassailable result.

A product of several excitedly interacting groups of computer scientists and mathematicians in the United States, Canada, and Israel, the idea of an interactive, zero-knowledge proof developed quickly. Initially, Shafi Goldwasser, Silvio Micali, and Charles Rackoff, motivated by theoretical questions concerning the efficiency of computer algorithms, worked out that it was, in principle, possible to convey that a theorem is proved without having to provide any details of the proof itself. Then Micali, Oded Goldreich, and Avi Wigderson demonstrated that zero-knowledge proofs can actually be found for certain theorems. Manuel Blum extended the scheme to cover any mathematical theorem. Amos Fiat and Adi Shamir, taking the idea into the world of spies and sensitive data, used it to formulate a secure, cryptographic method of identifying computer users.

Blum's scheme is interactive. It features a dialog between the prover, who has found a proof of a conjecture, and a skeptical verifier. The verifier can ask a question about the proof that requires the equivalent of a yes-or-no answer. If the prover really knows the proof, he can answer the question correctly every time it is posed. If he doesn't know the proof, the alleged prover has only a 50 percent chance of being right each time. After, say, a dozen tries, the chances of fooling the verifier get very small. Neither the question nor the possible answers give away even a hint of the proof itself—hence, the term zero-knowledge proof.

An example from graph theory illustrates how the scheme works. Any network of points, or nodes, connected by lines, or edges, is called a graph. In this case, the graph consists of a star-shaped pattern of lines linking 11 points (*see Figure* 8.1). The prover has found a continuous path along the

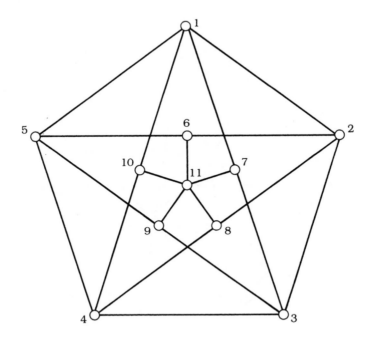

Figure 8.1 The Hamiltonian cycle in this star-shaped graph passes through points in the order 1, 5, 6, 2, 8, 4, 10, 11, 9, 3, 7, then back to 1.

connecting links that passes only once through each of the 11 points of the graph and returns to where it started. This special type of path is called a Hamiltonian cycle.

For the purposes of a zero-knowledge proof, the star-shaped graph in the example can be redrawn so that all the points fall on the circumference of an imaginary circle, with the connecting lines still joining the same points as in the original graph (*see Figure 8.2*). The prover's aim is to persuade a verifier that a Hamiltonian cycle is known without giving the verifier the slightest idea of how to construct the path. To do this, the prover privately marks 11 nodes along the circumference of a circle and labels them randomly from 1 to 11. The nodes on the circle are then connected in the same way as the points in the original graph. The resulting diagram is covered up by, say, an erasable opaque film like that used to hide words, icons, or numbers on certain lottery or contest tickets.

The verifier can ask the prover to uncover the complete graph, which shows that all the points are properly linked, or she can ask to see the Hamiltonian cycle. In the latter case, the prover erases enough of the film

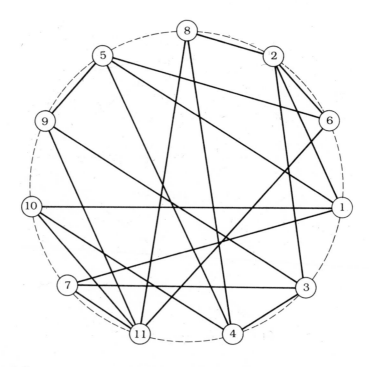

Figure 8.2 A star-shaped graph can be redrawn so that all the points fall on the circumference of an imaginary circle.

to reveal just the lines that make up the cycle. He can do this only if he knows the right path. However, because the nodes are still covered up, the verifier doesn't know the actual path from point to point. All she can verify is that a Hamiltonian cycle exists.

The process can be repeated as many times as the verifier wishes. Each time, the prover sets up a new circle diagram with randomly relabeled nodes, which is then hidden. Because he doesn't know whether the verifier will ask for the graph or the cycle, he has to be ready for either choice and therefore must know the cycle. Failure to produce either the correct graph or the cycle during a turn is equivalent to a wrong answer, and the verifier then knows that either the prover is lying or he doesn't have a correct proof.

Significantly, any mathematical theorem can be converted into a graph in such a way that if the theorem has a proof, the graph has a Hamiltonian cycle. Such a cycle is easy to find quickly with a proof in hand, and a known cycle can easily be turned into a proof. However, because the graphs may be incredibly complex jumbles of points and lines, simple inspection is rarely enough to identify a given graph's Hamiltonian cycle.

As Blum showed, a zero-knowledge scheme can handle any theorem provable within any logic system. The theorem prover proceeds by converting the proof to a graph with a known Hamiltonian cycle, and the rest follows. In the end, all a verifier finds out is that a theorem is provable and that the prover actually knows the proof.

As outlined, this scheme sounds somewhat cumbersome. However, the opaque film used in the example can be replaced on a computer by an encryption scheme that hides the information. Thus, proof checking can be done electronically, with the whole procedure encoded as strings of binary digits. It's also possible to make zero-knowledge proofs noninteractive. In a scheme developed by Blum and Micali, the prover can publish the fact that a theorem is true without revealing the proof and without the active participation of a verifier.

Such concepts can be used for password protocols and in cryptographic games, such as tossing a coin by telephone or exchanging secret keys, as outlined in Chapter 2. Indeed, it all sounds like a children's game, proving that you know a particular password without having to give away any hint of what the actual password is.

One such identification method, proposed by Uriel Feige, Amos Fiat, and Adi Shamir depends on the observation that while it's relatively easy to determine whether a number is prime, finding the factors of a large composite number is comparatively difficult and time-consuming. By following the appropriate mathematical procedure, the prover can use his knowledge of the prime factors of a large number to persuade a verifier, who knows the number but can't factor it, that his "signature" is valid.

Successfully implemented, methods involving interactive proofs could improve the security of computer operations such as transferring funds, signing contracts, and sending and receiving sensitive information. Neither eavesdropper nor recipient would be able to hijack enough information to masquerade as the sender.

Clear Thoughts

Suppose that after years of scrupulous reasoning and painstaking analysis, Peter Fermax of Enormous State University truly believes he has proved the infamous Snark Conjecture. Now he faces the formidable task of persuading his fellow mathematicians that his 1,210-page manuscript contains not a single error that would invalidate the proof. Who would have the time and patience to check each line and certify the proof's integrity?

Unlike his mathematical forebears, who typically had to struggle mightily to win acceptance of their lengthy proofs, Fermax has a tool at his disposal that greatly eases a proof checker's burden. Having already expressed his long chain of reasoning in an acceptable, logical form, he can call upon a supercomputer to convert his statements automatically into a holographic version of the same proof.

As in a laser-generated hologram, which makes possible the reconstruction of a scene not only from the whole recorded image but also from each tiny subsection of the image, every statement in a holographic proof contains information about the entire proof. Thus, an error in even a single line of the original proof has a high probability of showing up in any given line of the holographic proof. In effect, converting a proof into its holographic form greatly amplifies any mistakes present and makes them more readily detectable. To ascertain a proof's validity a checker has merely to examine, say, a dozen statements. There's no need to look at the whole proof—no matter how long.

Confident that his proof will withstand scrutiny, Fermax submits the holographic version to the editors of an electronic journal. Using a rudimentary desktop computer, an anonymous reviewer quickly checks the proof, and the Snark Conjecture becomes the Fermax Theorem.

Although such a scenario presently exists more in the realm of fantasy than in the real world, computer scientists have established the basic principles underlying this sort of scheme. Around 1990, Laszlo Babai of the University of Chicago, along with Lance Fortnow and Carsten Lund, established that although a small desktop computer can't keep up with a supercomputer's prodigious calculations, it can nonetheless interrogate a pair of such supercomputers independently working on the same problem and, within a reasonable time, determine whether the answers produced by the supercomputers are correct. In effect, they showed there is a mathematical way of conducting a session so that the supercomputers can be tricked into contradicting each other if there were an error present. That corresponds to the intuitive notion in a detective scenario that it's easier to get two suspects to contradict each other than it is to get a single suspect to trip up on a story.

Babai, Fortnow, Leonid Levin of Boston University, and Mario Szegedy of Bell Labs then improved on this procedure by noting that the interactive question-and-answer format of an interrogation is unnecessary. They devised a method for transforming a calculation or a mathematical proof into a so-called transparent (or holographic) form that magnifies any errors in the original. This technique involves writing the initial proof as a series of formal mathematical statements and "adding" them together in a special way. Mistakes in any of the original statements would show up in most of the resulting sums. The single, fragile thread of a traditional mathematical proof is thus replaced by the heavily redundant, multistranded cable of a holographic, or transparent, proof. Although the transparent proof, expressed in mathematical legalese, is a lengthy, rambling retelling of the original proof, it's one that's easy to see through if the prover lies or makes a mistake. A checker using a modest desktop computer could quickly verify the correctness of the transformed calculation or proof by examining only a relatively small number of these sums.

Other refinements of the method quickly followed, but the most surprising outcome—and perhaps the one of most immediate practical value—lay in the illumination of a deep, unexpected link between proof checking and the difficulty of solving certain kinds of problems in computer science. In some such cases, experience strongly suggests that any conceivable algorithm for finding the solution takes such a long time to run that the exact answer can't be reached within a reasonable amount of time when the number of choices grows large. The new results go one step further. They show that if it takes an unrealistically long time to compute the exact solutions of certain problems that involve choosing optimal strategies from a host of possibilities, finding acceptable approximate answers encounters the same barrier. For these cases, it's no easier to find approximate solutions than to find the exact answers. At this stage, however, it isn't clear how widespread this phenomenon is. The results obtained so far on the difficulty of arriving at approximate solutions apply to only a portion of the hard optimization problems that computer scientists face in handling everything from airline schedules to telephone networks.

At the same time, computer scientists and others are exploring what it would take to convert properly expressed mathematical proofs or computations into a transparent form. Anyone hoping to apply the technique to a real mathematical proof, written initially in a human language, would first have to find a way of translating it into a formal language based on the rules of logic. Another difficulty stems from the great length of the transparent proof that one would typically obtain from a given formalized proof. "We are trying to reduce that size, and once we reduce it to a reasonable level, one would hope that practical applications could be found," Babai says.

This doesn't mean that human mathematicians may become redundant at some stage. "We are not able to prove theorems. We are only able to check proofs," Babai warns. "This is an important distinction. To come up with a proof, you need ingenuity. To check a proof, you need only a machine."

Matters of State

One of the most provocative questions in mathematics is whether a mechanical method exists to determine the truth of any given mathematical statement. Is some kind of universal machine, following a strict procedure, capable of judging mathematical truth?

With one crucial qualification, the surprising answer is yes. Such an all-purpose machine not only is possible but also can do anything that a human mathematician or the most powerful computer conceivable in theory or practice can do. However, it can't answer mathematical questions that human mathematicians themselves can't answer using a mechanical procedure. For example, it can't determine all the digits of a nonrepeating decimal number, such as π, because any mechanical method would require an infinite number of steps.

The mathematician Alan Mathison Turing (1912–1954) was one of the first to propose the idea of a universal mathematics machine. Turing and Emil Leon Post (1897–1954) independently proved that determining the decidability of mathematical propositions is equivalent to asking what sorts of sequences of a finite number of symbols can be recognized by an abstract machine with a finite set of instructions. Such a mechanism, briefly mentioned in Chapter 7, is now known as a Turing machine. Given a large but finite amount of time, a Turing machine is capable of any computation that can be done by a modern digital computer, no matter how powerful. A supercomputer may do the job faster, but the slowpoke Turing machine also eventually reaches the finish line.

A Turing machine can be pictured as a black box capable of reading, printing, and erasing symbols on a single, long tape or strip of paper divided by lines into square cells, or boxes. Each cell is blank or contains one symbol from a finite alphabet of symbols.

The Turing machine scans the tape, one cell at a time, usually beginning at the cell furthest to the left that contains a symbol. The machine can leave the beginning symbol unchanged, erase it, or print another in its place. If in reading the tape the machine later encounters a blank cell, the machine has the choice of leaving the cell empty or entering a symbol. After it performs its assigned task on a given cell, the machine stops or moves one cell to the left or right.

What a machine does to a cell and which way it moves afterward depends on the state of the machine at that instant. Like a state of mind, the machine's internal configuration establishes the environment in which a decision is made. Turing machines are restricted to a finite number of states.

The most important part of a Turing machine is its action table, which is a bit like the software that instructs a computer; a Turing machine itself is somewhat like a typewriter whose characters can be put together in countless different ways. The same typewriter can produce any novel that has ever been written or ever will be written. Similarly, as Turing and Post were able to prove, a universal Turing machine can be programmed with a finite set of instructions to imitate any other special-purpose machine.

An action table stipulates what a machine will do for each possible combination of symbol and state. The first part of the instruction specifies what the machine should write, if anything, depending on which symbol the machine sees. The second part specifies whether the machine is to shift one frame to the left or to the right along the tape. The third part determines whether the machine stays in the same state or shifts to another state, which usually has a different set of instructions.

Suppose a Turing machine must add two integers. There are numerous different ways in which this can be done, depending on how many symbols and states are allowed. One of the simplest possibilities is to represent an integer by a string of asterisks. Thus, *** would be 3 and **** would be 4. To add 3 and 4, *** and **** are first printed on a tape, with a blank space between the two strings. The machine then fills in the blank cell by printing an asterisk and goes to the end of the string of asterisks and erases the last one in the row. What's left is the required answer—a string of seven asterisks.

An action table (below) instructs the machine how to perform the addition. The table's first column gives the machine's possible mental states, and the first row lists all the symbols being used. In this example, there are only two symbols: an asterisk (*) and a blank. Each combination of symbol and state specifies what, if anything, needs to be done to a cell, in which direction to move after the action, and the state of the machine, that is, which set of instructions it will follow for its next move.

Symbol	*	Blank
State 0	Right, state 0	Print *, right, state 1
State 1	Right, state 1	Left, state 2
State 2	Erase, stop	—

The machine begins in state 0 and scans the asterisk farthest to the left on the tape. According to the instructions for state 0 and symbol *, the machine leaves the symbol as it is, shifts one step to the right, and remains in state 0. It encounters another asterisk, and the process is repeated. Finally, it reaches the blank cell. From the table, the machine knows it must print an asterisk, then move one space to the right. This time, however, it shifts into a new state. Now the machine stays in state 1 until, step by step, it reaches the first blank at the end of the string. This time, it backs up one space and shifts into state 2. It erases the cell's asterisk and stops. The addition is complete (*see* Figure 8.3).

The same action table can generate the sum of any two whole numbers, no matter what their size. However, adding two numbers such as 49,985 and 51,664 would require a tape with at least 100,000 cells. To be capable of adding any two numbers, the tape would have to be infinitely long. In fact, a universal Turing machine capable of any mathematical operation, must have an infinitely long tape. An ordinary computer, with a limited amount of memory, lacks this property.

Similar tables can be worked out for subtraction and for practically any other mathematical operation. The sole condition is that the number of states and symbols listed in such a table is finite, which ensures that a routine, mechanical process can do the job. Often, several different tables can be found to perform the same operation.

The world of Turing machines has countless nooks and crannies worth exploring. One such corner is occupied by a particular group of Turing machines dubbed "busy beavers." These machines, following an action table with a strictly limited number of states, must print as many symbols (say, the digit 1) as possible on an initially blank tape before they grind to a halt. The champion machine for a given number of states—the busy beaver—is the one that prints the maximum number of 1s.

Mathematicians already know that a three-state busy beaver writes at most six 1s before it comes to a halt (*see* Figure 8.4). The four-state busy beaver stops after writing 13 1s. The answer for a five-state busy beaver isn't known yet, but several researchers have managed to push the number up to at least 4,000.

State

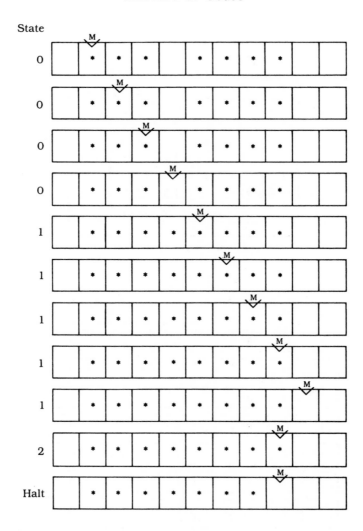

Figure 8.3 At each step, a Turing machine may move one space to the left or right. By following a simple set of rules, the machine can add two numbers, starting with separate groups of three and four asterisks and ending up with one group of seven asterisks.

In 1984 the amateur mathematician George Uhing built a small computer that automatically tested, one after another, a selection of the 64,403,380,965,376 different possible five-state, two-symbol (1 and 0) Turing machines. Many of the candidate machines were easy to eliminate because their action tables lead them into infinite loops, and the machines would run on forever. Other half-hearted contenders stop almost immediately.

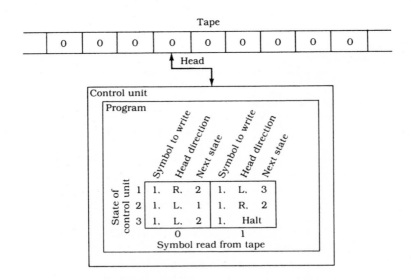

Tape

| 0 | 0 | 0 | 0 | 0 | 0 | 0 | 0 | 0 |

Head

Control unit

Program

State of control unit		Symbol to write	Head direction	Next state	Symbol to write	Head direction	Next state
	1	1,	R,	2	1,	L,	3
	2	1,	L,	1	1,	R,	2
	3	1,	L,	2	1,	Halt	
		0			1		

Symbol read from tape

State	Tape								
1	0	0	0	0	0	0	0	0	0
2	0	0	0	0	1	0	0	0	0
1	0	0	0	0	1	1	0	0	0
3	0	0	0	0	1	1	0	0	0
2	0	0	0	1	1	1	0	0	0
1	0	0	1	1	1	1	0	0	0
2	0	1	1	1	1	1	0	0	0
2	0	1	1	1	1	1	0	0	0
2	0	1	1	1	1	1	0	0	0
2	0	1	1	1	1	1	0	0	0
2	0	1	1	1	1	1	0	0	0
1	0	1	1	1	1	1	1	0	0
3	0	1	1	1	1	1	1	0	0
Halt	0	1	1	1	1	1	1	0	0

Figure 8.4 A three-state busy beaver Turing machine, starting with a blank tape, writes a string of six 1s before it comes to a halt. A simple program tells the machine what to do at each step, depending on whether the machines reads a 1 or a 0.

After letting his computer run for about three weeks while it sifted through several million possibilities, Uhing found a five-state Turing machine that prints out 1,915 1s, making more than 2 million moves to generate the string. The string itself consists of alternating 1s and 0s, except for the last seven spaces, where the sequence ends with 110 1111.

In 1989, Heiner Marxen of the Technical University of Berlin topped Uhing's mark by finding a five-state Turing machine that prints 4,098 1s on a tape, taking nearly 12 million steps. Whether this particular machine is the long-sought ultimate busy beaver isn't known. Other sets of five instructions, not yet identified, may conceivably print out even more 1s before coming to a stop. Marxen's solution already requires so many moves that it would probably take an army of people, using the fastest available computers, many years to search for better machines. His result implies that the behavior of even a simple digital machine can quickly get out of hand, much sooner than mathematicians had expected. Translated into mathematical terms, the large jump evident in the output from a four-state to a five-state busy beaver means that the amount of computation, or the number of steps, needed to solve particular problems can easily outstrip the problem-solving capacity of real computers.

Moreover, determining whether a particular Turing machine stops is tantamount to proving or disproving certain mathematical conjectures. Marxen's result illuminates the difficulty of distinguishing between machines that halt and machines that don't halt. The difference between a decidable and an undecidable proposition may be as subtle as the difference between a machine that runs on forever and one that goes on for so long that everyone loses patience waiting for it to stop.

What is remarkable about Turing machines in general is that these simple devices embody a method of mathematical reasoning. Moreover, a Turing machine doesn't have to be a mechanical device. It can be a computer, a mathematician, a team of students, a game, or any other entity that follows some sort of action table.

Interestingly, John Conway was able to prove that the patterns observed in the game of "Life" (see Chapter 7) can function as universal computers; in other words, the game has patterns suitable for the construction of a Turing machine. In fact, the logic circuitry of any possible computer can be built from a set of four patterns that appear in the game—guns, gliders, eaters, and blocks.

Going one step further, the "Life" universe itself can be thought of as an array of computers or a single massively parallel computer. The same rules are applied to each cell, and the "Life" universe accepts data in the form of a particular starting pattern of marked or living cells. After the rules are applied, the new pattern of living cells comprises the output data. In this way, the infinite plane on which "Life" is played supplies the answers to mathematical questions.

There's also an intriguing link between Turing machines and chaotic dynamical systems (*see Chapter 6*). To predict the future course of a dynamical system, such as a swinging pendulum, physicists have traditionally counted on solving a suitable equation to find a formula describing the system's position or state at any time. However, many of the equations used to model physical systems do not yield such formulas. In this situation, researchers turn to methods that provide approximate answers, which usually offer only a limited degree of predictability.

Indeed, the course of chaotic dynamical systems depends so sensitively on initial conditions that a prediction's accuracy is strictly limited by the amount of information available. The farther into the future one wants to predict, the more precisely one needs to know the initial conditions. For some dynamical systems, however, the future seems even more clouded. In 1990, Christopher Moore, then a graduate student at Cornell University, constructed a mathematical scheme corresponding to dynamical systems that are, in a sense, more unpredictable than chaotic systems. Even if the initial conditions are known exactly, virtually any question about their long-term dynamics is undecidable.

One approach to studying dynamical systems involves converting a differential equation or some other complicated mathematical expression into a kind of game played with specific rules for manipulating sequences of symbols or digits—often just 0s and 1s. Moore turned that idea on its head by first inventing a new type of game with interesting characteristics, then searching for connections with dynamical systems.

Moore was able to show that his games, which he calls generalized shifts, correspond to dynamical systems known as iterated maps, which researchers sometimes use to model physical systems. An iterated map represents the results of substituting each of a set of initial values into a mathematical expression, calculating the answer, then substituting that answer back into the original expression to calculate a new answer, and so on.

Moore also demonstrated that, in certain cases, his games are equivalent to Turing machines. Various properties of Turing machines can therefore be transferred to iterated mappings, so the maps inherit all the sorts of complexities of Turing machines—complexities very different from those that one generally has in chaos.

A properly constructed Turing machine, operating under rules that specify its step-by-step behavior, can perform any computation that a computer can do. However, a particular Turing machine might be saddled with a problem that takes an inordinate—perhaps infinite—amount of time to solve. For a given initial state, such a machine may never stop once it gets going, and therefore may never come up with an answer.

Thus, chaotic behavior is unpredictable because the description of its initial conditions is bound to contain small uncertainties, and these errors grow until the prediction is completely off. Turing machines are unpredictable even if the initial conditions are known exactly. Consequently, if a step in an iterated mapping happens to correspond to a generalized shift that incorporates a Turing machine saddled with an "impossible" problem, researchers would have practically no way of determining what ultimately results from those particular initial conditions. The best one can do is to simulate the system and see what happens. Even the simplest long-term properties of the motion are undecidable.

Researchers studying physical systems that happen to correspond to such a mathematical scheme would have no way of determining the system's future, not just in the long term but also in the short run. Moreover, the system would behave so irregularly that reliable, statistical measures of its average or overall behavior would be impossible to obtain.

No one yet knows whether this type of dynamics arises in realistic physical systems. Nonetheless, although the dynamical systems Moore studied appear contrived, they suggest that it may be worth looking for a similar level of complexity in, say, the so-called three-body problem, which concerns the complicated motions of three gravitationally interacting objects.

Automatic Logic

Paul Erdös (1913–1996) liked to describe a mathematician as a "device for turning coffee into theorems." To Erdös, however, it wasn't enough to cobble together any old proof to establish the truth of a conjecture. He subscribed to the notion that God has a book containing all the theorems of mathematics, together with their most elegant and beautiful proofs. The goal of mathematicians was to uncover these sublime instances of logical reasoning. When Erdös wanted to express particular appreciation of a proof, he would exclaim, "This is one from the book!"

One result unlikely to qualify for "the book" is a recent proof of the so-called Robbins conjecture, which concerns an aspect of the basic rules of logic. Nonetheless, the proof is noteworthy because it was found by a computer program, succeeding spectacularly where mathematicians had failed. Originally proposed in the 1930s by Herbert Robbins, the problem had stumped everyone who tackled it over the years. It finally succumbed to a computer program designed to reason in a general way rather than a program designed to solve a particular problem, whether an equation or the moves of a chess game. Called EQP for "equational prover," the automated reasoning software was developed by William McCune, a computer scientist at the Argonne National Laboratory.

Computers are not new to the business of proving theorems. In the past, however, they generally served as bookkeepers or assistants to check the large number of special cases needed to establish a theorem (*see Chapter 1*). Mathematicians or computer scientists outlined the necessary steps, wrote the special-purpose software, and specified the calculations required for a proof. McCune's program does something different. Although the computer's search has constraints, neither the path to a result nor the ultimate outcome is determined beforehand. Indeed, because typically no one knows any route to a solution, the search can't be guided in a meaningful way.

The development of software for automated reasoning began at Argonne in the 1960s with the work of the mathematician Larry Wos and his colleagues. They focused on how to describe a mathematical problem in terms that a computer can handle and how to get it to prove theorems by drawing conclusions that follow inevitably and logically from given postulates, or axioms. However, they never tried to imitate what people do.

The idea was to give an automated reasoning system the statement of a conjecture along with a few rules of thumb constraining the search among the infinite number of possible logical paths to a proof, then allow the system to proceed on its own. Advances came in the form of improved strategies for keeping the computer from getting entangled in lengthy chains of reasoning that apparently led nowhere. Those improvements, along with faster computers and better software engineering, eventually produced an agile reasoning system called Otter, aimed at questions in abstract algebra and formal logic.

In the 1990s, Wos, McCune, and their coworkers used Otter to solve a wide variety of mathematical problems, including original proofs of theorems in logic, algebraic geometry, group theory, and other areas of mathematics. Then McCune decided to make a fresh start and develop a daughter of Otter, incorporating techniques and strategies that researchers had found useful while experimenting with Otter during the previous decade. The result was EQP. The program featured a number of strategies for selecting possible search paths, including simply following the path defined by the smallest equation or formula at certain stages or side-stepping expressions containing more than, say, 100 logical relationships.

Initial tests of EQP were encouraging, and the program succeeded in proving, sometimes for the first time, several theorems of special interest to a few logicians and other experts. These results, however, were of limited value to most mathematicians. As a benchmark test of his system, McCune turned to the Robbins conjecture, which states that a set of three equations in logic is equivalent to a Boolean algebra.

Boolean algebra is a mathematical model of some of the basic rules of logic. It includes such laws as "for any proposition P, the negation of the negation of P means the same thing as P" or "for any two propositions, P, Q, the conjunction of P and Q is false if and only if one or both of them is false." The Robbins problem is equivalent to proving that the equation not(not(P)) = P can be derived from the following three equations:

$$P \text{ or } Q = Q \text{ or } P$$

$$(P \text{ or } Q) \text{ or } R = P \text{ or } (Q \text{ or } R)$$

$$\text{not(not(P or Q) or not(P or not(P or not(Q)))) = P}$$

In other words, the problem is to determine whether this particular set of equations incorporates the basic laws of Boolean logic, which specify the different ways in which collections of objects can be combined or manipulated using logic operations such as "and," "or," and "not." Such logic underlies the workings of today's digital computers and is often used to facilitate database searches.

Robbins himself worked on the problem and it was later taken up by others, including the prominent logician Alfred Tarski (1901–1983). When Tarski failed to make any headway, he handed the problem out as a challenge to graduate students and visitors. Wos first came across the problem in 1979, when he helped a student develop a novel way of attacking it by working backward—that is, by finding conditions that, if true, would prove the theorem. That promising start, however, did not lead to an immediate proof.

At the same time, the problem had characteristics ideal for an automated attack. Embodied in just three equations, the Robbins conjecture could be stated succinctly in a form a computer could understand. Moreover, researchers could identify a number of conditions that, if true, would each point to a complete proof. In the fall of 1996, McCune's EQP was ready for a new challenge, and McCune let the prover software grapple with the Robbins conjecture. The computer started to work, and McCune checked it every day for progress. On the eighth day, the computer found an answer and stopped. It had proved the Robbins conjecture is true.

Fresh out of the computer, the proof consisted of a cryptic chain of lengthy, practically unreadable statements of logic. McCune then refined EQP's procedures to obtain somewhat more streamlined proofs. In the meantime, several mathematicians independently checked the computer proof and found it correct. Stanley N. Burris of the University of Waterloo, for one, converted the proof's long statements into simpler forms that could be more readily understood by humans. "I ended up with a lot of little equations," he noted. "You could easily sit on a bus and go through the hundred or so steps of the proof."

Branden E. Fitelson , a philosophy graduate student at the University of Wisconsin-Madison also worked through the result. "It's a very complicated proof, not at all elegant or conceptual," he remarked. "It's a perfect example of the type of proof that a machine can find but we can't because of its complexity."

EQP's success raises some interesting questions about mathematical creativity. If a mathematician had proved the conjecture, the achievement would have been heralded as a significant accomplishment, but it's highly unlikely that a mathematician would have come up with the kind of proof that EQP found.

At the same time, mathematicians generally have no way of telling beforehand, just by looking at a problem or a conjecture, what kind of attack will subdue it. In fact, one of the attractions of the Robbins conjecture was that it looked vulnerable to techniques mathematicians had used to prove similar theorems, differing only in what seemed like minor details.

No one knows whether the computer proof generated by EQP represents a fluke success or the first of many triumphs. That's something that can be judged only in hindsight. Although mathematicians and computer scientists have developed demonstrably strong theorem provers, they don't yet know what power is required to be really competitive with human beings.

To see what lessons could be learned, McCune, Wos, and others took a close look at EQP's performance. "What was it doing during those eight days?" Wos wondered. "Is there some kind of strategy that would have tightened up its reasoning?" In the long term, the researchers are also interested in extending the kind of deductive reasoning an automated theorem prover can apply from relatively simple logical relationships involving equality to more complicated expressions. They would like to introduce mathematical induction, which would allow the computer to make inferences.

"Our programs do not learn, do not make judgments, and they do not invent concepts," Wos noted. Nonetheless, it may be possible to program them to be self-analytical—to recognize in a mechanical sense when searches are not going well and to shift strategies appropriately.

With automated reasoning systems, Wos and McCune envision offering mathematicians the option of using a computer to free them of the drudgery of finding proofs and to allow them to spend more time on harder, more interesting problems. Programs such as Otter and EQP already work well in specialized areas of pure mathematics, such as the theory of algebraic systems. It's possible that computer-generated proofs will represent a significant influence in the near future in those fields. Similar approaches could help engineers design circuits or allow computer programmers to verify the correctness of their software. Indeed, research groups throughout the world are working on such applications of automated reasoning.

Computers have yet to come up with a proof that belongs in "the book," and they can't turn coffee into theorems, but the power of theorem-proving programs continues to increase. "Can machines reason at the same level as humans?" Burris asks. "In the long run, I would bet on it, but it's unlikely I'll see it in my lifetime."

———————— Burden of Proof ————————

Like other sciences, mathematics has an experimental component. In the trial-and-error process of developing and proving conjectures, mathematicians collect data and look for patterns and trends. They construct novel forms, seek logical arguments to strengthen a case, and search for counterexamples to destroy an argument or expose an error. In these and other ways, mathematicians act as experimentalists, constantly testing their ideas and methods.

The concept of proof, however, brings something to mathematics that is missing from other sciences. Once the experimental work is done, mathematicians have ways to build a logical argument that pins either the label "true" or "false" on practically any conjecture. Physicists can get away with overwhelming evidence to support a theory. In mathematics, a single counterexample is enough to sink a beautiful conjecture. For this reason, too, no set of pretty pictures, collected during countless computer experiments, is complete enough to substitute for a mathematical proof.

Emphasis on proof engenders both a strong conservatism and a deep skepticism among mathematicians. Their professional fame is tied to mathematical proofs of significant conjectures, and the rush to be the first with a correct proof adds considerable drama—and confusion, at times. Two recent examples illustrate the pitfalls strewn across a mathematician's path.

In March 1986 the British mathematician Colin Rourke and his student Edward Rêgo, from Portugal, announced that they had proved the Poincaré conjecture. To mathematicians, especially topologists, proving the Poincaré conjecture was something like being the first to climb Mount Everest. For more than 80 years, numerous mathematicians had stumbled over this famous problem, always slipping somewhere along the way. Sometimes, only a tiny gap—a subtle error buried within pages of mathematics—had halted the ascent.

As discussed in Chapter 4, the Poincaré conjecture proposes that any object that mathematically behaves like a three-dimensional sphere is topologically a three-dimensional sphere, no matter how distorted or twisted its shape may be. Although the statement sounds obvious, the difficulty lies in the enumeration of all the different ways in which three-dimensional space can be stretched and molded to form geometric surfaces called manifolds.

Because so many false "proofs" of the Poincaré conjecture had been proffered in the past, mathematicians approached the new claim warily. There was considerable skepticism. Rourke and Rêgo's approach to proving the conjecture was similar to ideas that had been tried unsuccessfully years before. The biggest problem in verifying the proof, which ran to dozens of pages in manuscript form, was to find mathematicians willing to take the time from their own work. If a mistake happened to be subtle, the task of checking the proof could take years of effort.

Rourke was so sure that the proof was legitimate that he took an unusual step for a mathematician and announced the results in a press release and, later in 1986, wrote an article about the proof for a popular science magazine. Meanwhile, most mathematicians keenly interested in the question remained silent and simply waited for better evidence. Some resented the fact that Rourke had publicized his achievement without waiting for verification. Normally, mathematicians quietly pass their manuscripts to their friends and colleagues for comment. Only months or years later, after the mathematical community has finally applied its stamp of approval, is the work published in a mathematical journal, and the public may then hear of the result.

In November of the same year, Rourke was at the University of California at Berkeley, conducting a seminar to explain and defend his proof. By the end of the week, Rourke's audience, which included some of the world's top topologists, had pointed out a gap in his proof, one that Rourke could not fill. In the end, there was no valid proof.

In general, the mathematical community likes to keep its disputes within the family and sweep messy details away. It's not surprising that outsiders rarely see mathematics as the exciting, human endeavor that it is. What emerges instead is an elegant, remote architecture, with the scaffolding down and the blueprints stored away. The human element is hidden.

The second story has a happier ending, and it involves one of the most celebrated conjectures in mathematical history. The problem started out as a note scrawled in the margin of a book. Now known as Fermat's last theorem, it was first proposed by the seventeenth-century French jurist and mathematician Pierre de Fermat. In a tantalizing sentence that was to haunt (and taunt) mathematicians for centuries to come, Fermat noted that although he had a remarkable proof of the theorem, he didn't have enough room to write it out. After Fermat died, scholars could find no trace of the proof in any of his papers.

Fermat's conjecture is related to an observation by the third-century Greek mathematician Diophantus of Alexandria that there are positive integers, x, y, and z, which satisfy the equation $x^2 + y^2 = z^2$. (For example, $3^2 + 4^2 = 5^2$.) In fact, this equation has an infinite number of positive-integer solutions. Fermat proposed that there are no such solutions to the equation $x^n + y^n = z^n$, when n is a whole number greater than 2.

A century after Fermat's death in 1665, Leonhard Euler proved that Fermat's conjecture is true for $n = 3$. Mathematicians later found proofs for other special cases, and computer searches performed in recent years showed that Fermat's last theorem is true for all exponents less than 4 million. Therefore, if a counterexample exists, it would involve huge numbers, but a proof of the general case remained elusive.

One effort to tame Fermat's conjecture started with the work of Gerhard Frey in the mid-1980s, who happened to be looking at equations of elliptic curves—equations generally written in the form $y^2 = x^3 + ax^2 + bx + c$, where a, b, and c are constants (*see Figure* 8.5). He found a way to express Fermat's last theorem as a conjecture about elliptic curves. That brought Fermat's problem into an area of mathematics for which mathematicians had already developed a wide range of techniques for solving problems.

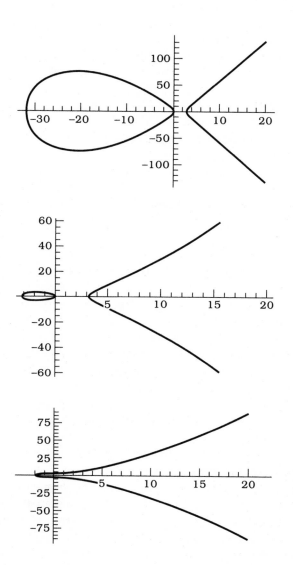

Figure 8.5 Examples of elliptic curves. *Top:* $y^2 = x(x - 3)(x + 32)$; *middle,* $y^2 = x^3 - 10x$; *bottom,* $y^2 = x^3 + 7$.

Frey wrote down the elliptic curve that would result if Fermat's conjecture is false. The curve turned out to have peculiar properties, and mathematicians examining the weird curve's equation had the feeling that the curve could not exist. If that could be proved, Fermat's conjecture would have to be true. Frey started on this task, but he failed to complete the proof.

The French mathematician Jean-Pierre Serre then suggested that one of his own conjectures, if proved, would help patch up Frey's effort. In 1987, Kenneth A. Ribet worked out the necessary proof of Serre's conjecture for a large class of situations. Ribet ended up showing that Fermat's purported theorem is true if certain elliptic curves arise from so-called cusp forms.

In 1988 the Japanese mathematician Yoichi Miyaoka proposed a proof for a key link in an alternative chain of reasoning that also led to Fermat's last theorem. His method, building on work done by several Russian mathematicians, connected important ideas in three mathematical fields—number theory, algebra, and geometry. Although highly technical, his argument filled fewer than a dozen manuscript pages, which is short for such a significant modern proof.

However, careful scrutiny of Miyaoka's complex proof turned up several flaws that cast doubt on the proof's validity. There was no straightforward way around the difficulties. Changing some statements in one part of the proof forced changes in many other parts, pulling apart the carefully constructed argument.

Meanwhile, Andrew Wiles, a young mathematician at Princeton University, had started to work in virtual secrecy on aspects of number theory closely related to Fermat's last theorem. Unlike Miyaoka's approach, the attack chosen by Wiles relied on the recently discovered links between Fermat's conjecture and the theory of elliptic curves. By assuming that Fermat's last theorem is false, mathematicians could construct a weird elliptic curve that they believe, for other mathematical reasons, shouldn't exist. Moreover, the existence of the strange curve would also contradict the so-called Taniyama-Shimura conjecture, which involves other characteristics of elliptics. Hence, if proving certain aspects of the Taniyama-Shimura conjecture excluded the strange curve, this would establish the validity of Fermat's last theorem.

Wiles took on the extremely challenging, highly technical task of proving the Taniyama-Shimura conjecture itself. However, he proceeded in such secrecy that even his closest acquaintances and colleagues were unaware of the extent of his determined effort. When Wiles was finally ready to reveal that he had proved a significant part of the Taniyama-Shimura conjecture, he chose to describe his results in his native land during three lectures presented at a workshop at the Isaac Newton Institute for Mathematical Sciences of the University of Cambridge in England, the university where Wiles had done his doctoral studies. His audience included Ribet, Barry Mazur of Harvard University, and many other experts in this particular specialty.

At the end of his third lecture, almost as an afterthought, Wiles noted that he had proved enough of the Taniyama-Shimura conjecture to show that Fermat's last theorem was true. His dramatic announcement on June 23, 1993, caught the mathematical community completely by surprise.

By following a course that built on previous, well-understood results, and because of his own reputation for being extremely cautious and careful in his mathematical work, Wiles immediately earned a great deal of respect for his proof. Nonetheless, the details of Wiles's proof, running to more than 200 manuscript pages, had to be checked thoroughly by experts. There was always a chance that there was some subtle flaw, even though Wiles had used well-known fundamental theorems and his basic approach appeared correct.

Wiles submitted a preliminary manuscript of the proof to the journal *Inventiones Mathematicae*. Copies of the manuscript were also sent to about half a dozen mathematicians for checking, but no copies were circulated publicly. Wiles himself continued to work quietly on his proof, clarifying arguments and correcting minor errors pointed out by the referees of his paper.

That fall, the process of smoothing out and filling in the technical details of Wiles's celebrated result turned up a troublesome gap in the proof's logic. In December, Wiles sent out an electronic-mail message to colleagues acknowledging that reviewers of the proof had pointed out a number of problems, one of which remained unresolved. His argument foundered near the end, where Wiles had relied on a powerful new method developed by the Russian mathematician Viktor A. Kolyvagin to serve as a kind of mathematical bookkeeping system.

Disappointed but intent on continuing, Wiles said at the time, "I believe that I will be able to finish this in the near future using the ideas explained in my Cambridge lectures." In February 1994, Wiles began giving a detailed account of his work in a course he was teaching at Princeton, covering parts of the proof that had already been checked and verified. Meanwhile, he refined the first few chapters of his lengthy manuscript into a form suitable for circulation.

Wiles also expended a great deal of effort thinking about the obstacle lying in his path toward Fermat's last theorem. In a rare public lecture in May 1994, a visibly relaxed Wiles explained that he believed the obstacle was probably surmountable—that he understood the problem well enough to see its connection with something more standard in mathematics.

At the same time, despite the gap and the complexity of the approach, Wiles's work quickly inspired several mathematical efforts on related questions. "Any time there's a major achievement like this, it changes what people work on and how they think about things," Ribet remarked. Indeed, mathematicians have new techniques—developed by Wiles—for tackling other important, difficult questions in number theory. For the specialist, Wiles's work on Fermat's last theorem is the less exciting part and heralds the beginning of a new subject.

By November 1994, Wiles was certain that he had bridged the gap to complete his proof. He began distributing copies of two new manuscripts that addressed the concerns raised about his original argument. The first, a lengthy paper, announced the revised proof, still following closely the strategy Wiles had outlined in his Cambridge lectures more than a year earlier. The second, a short paper written in collaboration with the Cambridge mathematician Richard L. Taylor, contained mathematical reasoning justifying a key step in the main proof.

To circumvent the original problem, Wiles had decided to go back to an approach he had tried several years earlier but had abandoned. Helped by Taylor, who had come to Princeton to work on the problem, Wiles succeeded in establishing the conditions he needed to complete the proof. The new approach turned out to be significantly simpler and shorter than the original one. Wiles's proof was now solid and his triumph complete.

Although the final proof of an important conjecture may be a great achievement, it generally rests on the work—the ideas, conjectures, and techniques—of many mathematicians. The communal building process emphasizes again the importance of proofs for establishing a firm foundation on which future generations of mathematicians can build new structures.

Selecting the appropriate problem to tackle is one of the arts a mathematician must master to succeed in the field. Paul Erdös once noted: "A well chosen problem can isolate an essential difficulty in a particular area, serving as a benchmark against which progress in this area can be measured. An innocent looking problem often gives no hint as to its true nature. It might be like a 'marshmallow,' serving as a tasty tidbit supplying a few moments of fleeting enjoyment. Or it might be like an 'acorn,' requiring deep and subtle new insights from which a mighty oak can develop."

Judging the difficulty of a problem can be a tricky matter, however. In the early 1920s, the great mathematician David Hilbert presented a lecture in which he said that he expected to see within his lifetime a proof of the famous Riemann hypothesis, which concerns a surprising link between number theory and the calculus of complex numbers. He predicted that although he probably wouldn't see a proof of Fermat's last theorem within his lifetime, others who outlast him might. He asserted that no one will see a proof that $2^{\sqrt{2}}$ is a so-called transcendental number.

Hilbert was right (just barely) on Fermat's last theorem. The Riemann hypothesis, however, remains unsolved. And the transcendence of $2^{\sqrt{2}}$ was proved just a few years after Hilbert's lecture, and that proof rested on deep mathematical results obtained by Carl Ludwig Siegel (1896–1981), who was in the audience for the lecture. There is a deep, enduring element of unpredictability in the conduct of mathematical research.

———————— —— — — ———— ——

Mathematics is an exciting, dynamic field that continues to generate provocative ideas and novel concepts. Some notions quickly find their way into applications; others rest on their intrinsic beauty; still others occupy a modest place in the growing structure of mathematics itself. As the mathematician Rick Norwood eloquently noted in his essay "In Abstract Terrain":

> So it goes, new mathematics from old, curving back, folding and unfolding, old ideas in new guises, new theorems illuminating old problems. Doing mathematics is like wandering through a new countryside. We see a beautiful valley below us, but the way down is too steep, and so we take another path, which leads us far afield, until, by a sudden and unexpected turning, we find ourselves walking in the valley.

For the mathematical tourist, a trek across abstract terrain can be both revealing and rewarding. Peeking into the minds of mathematicians, reveling in the intricacies of number and shape, and tangling with intriguing foreign concepts provide much food for deep thought. Like a physician examining the dark and light areas of an X-ray image, the tourist glimpses the mathematical skeletons that underlie our understanding of the everyday world and begins to appreciate the wonders and mysteries of mathematics.

FURTHER READING

GENERAL WORKS

Boyer, Carl. 1985. A *history of mathematics*. Princeton, N.J.: Princeton University Press.

Cipra, Barry. 1993, 1994, 1996. *What's happening in the mathematical sciences*, 3 vols. Providence, R. I.: American Mathematical Society.

Courant, Richard, Herbert Robbins, and Ian Stewart. 1996. *What is mathematics? An elementary approach to ideas and methods*, 2nd ed. New York: Oxford University Press.

Davis, Donald M. 1993. *The nature and power of mathematics*. Princeton, N.J.: Princeton University Press.

Devlin, Keith. 1994. *Mathematics: The science of patterns*. New York: Scientific American Library.

Peterson, Ivars. 1990. *Islands of truth: A mathematical mystery cruise*. New York: W. H. Freeman.

Stewart, Ian. 1996. *From here to infinity: A guide to today's mathematics*. New York: Oxford University Press.

1 EXPLORATIONS

Appel, Kenneth, and Wolfgang Haken. 1986. The four color proof suffices. *Mathematical Intelligencer* 8, no. 1: 10–20.

———. 1977. The solution of the four-color-map problem. *Scientific American* 237 (October): 108–121.

Barnette, David. 1983. *Map coloring, polyhedra, and the four-color problem*. Washington, D.C.: Mathematical Association of America.

Bracewell, R. N. 1990. Numerical transforms. *Science* 246 (May 11): 697–704.

Cipra, Barry A. 1996. Advances in map coloring: Complexity and simplicity. *SIAM News* (December): 20.

Dewdney, A. K. 1990. How to resurrect a cat from its grin. *Scientific American* 263 (September): 174–177.

Diaconis, Persi, and R. L. Graham. The Radon transform on $\mathbf{Z}_2^k$. *Pacific Journal of Mathematics* 118 (June): 323–345.

Gardner, Martin. 1977. Extraordinary nonperiodic tiling that enriches the theory of tiles. *Scientific American* 236 (January): 110–122.

———. 1966. The four-color map theorem. In *Martin Gardner's new mathematical diversions from Scientific American*. New York: Simon and Schuster.

Grünbaum, Branko, and G. C. Shephard. 1987. *Tilings and patterns*. New York: W. H. Freeman.

Jensen, Tommy R., and Bjarne Toft. 1995. *Graph coloring patterns*. New York: Wiley.

Nelson, David R. 1986. Quasicrystals. *Scientific American* 255 (August): 42–51.

Norwood, Rick. 1982. In abstract terrain. *The Sciences* 22 (December): 13–18.

Olivastro, Dominic. 1992. Color proof. *The Sciences* 32 (May/June): 53–55.

Peterson, Ivars, 1996. Clusters and decagons. *Science News* 150 (October 12): 232–233.

_____. 1991. Shadows and symmetries. *Science News* 140 (December 21/28): 408–410.

_____. 1990. Tilings for picture-perfect quasicrystals. *Science News* 137 (January 13): 22.

_____. 1989. The color of geometry. *Science News* 136 (December 23/30): 408–410, 415.

_____. 1988. Tiling to infinity. *Science News* 134 (July 16): 42.

_____. 1988. Shareware, mathematics style. *Science News* 133 (January 2): 12–13.

_____. 1986. Heart flow. *Science News* 130 (September 27): 204–206.

_____. 1986. Inside averages. *Science News* 129 (May 10): 300–301.

_____. 1985. The fivefold way for crystals. *Science News* 127 (March 23): 188–189.

_____. 1985. Quasicrystals: A new ordered structure. *Science News* 127 (January 19): 37.

Robertson, Neil, et al. A new proof of the four-colour theorem. *Electronic Research Announcements of the American Mathematical Society* 2:17–25.

Senechal, Marjorie. 1995. *Quasicrystals and geometry.* Cambridge: Cambridge University Press.

Shepp, L. A., and J. B. Kruskal. 1978. Computerized tomography: The new medical X-ray technology. *American Mathematical Monthly* 85 (June/July): 420–439.

Steinhardt, Paul Joseph. 1986. Quasicrystals. *American Scientist* 74 (November-December): 586–597.

Tymoczko, Thomas. 1979. The four-color problem and its philosophical significance. *Journal of Philosophy* 76 (February): 57–83.

2 PRIME PURSUITS

Apostol, Tom M. 1996. A centennial history of the prime number theorem. *Engineering & Science* 59, no. 4: 18–28.

Blair, W. D., C. B. LaCampagne, and J. L. Selfridge. 1986. Factoring large numbers on a pocket calculator. *American Mathematical Monthly* 93 (December): 802–808.

DeMillo, Richard, and Michael Merritt. 1983. Protocols for data security. *Computer* (February): 39–50.

Diffie, Whitfield, and Martin E. Hellman. 1979. Privacy and authentication: An introduction to cryptography. *Proceedings of the IEEE* 67 (March): 397–427.

Gardner, Martin. 1964. The remarkable lore of the prime numbers. *Scientific American* 210 (March): 120–128.

Garfinkel, Simson L. 1996. Public key cryptography. *Computer* (June): 101–104.

Golomb, Solomon W. 1985. The invincible primes. *The Sciences* 25 (March/April): 50–57.

Further Reading

Guiasu, Silviu. 1995. Is there any regularity in the distribution of prime numbers at the beginning of positive integers? *Mathematics Magazine* 68 (April): 110–121.

Hawkins, David. 1958. Mathematical sieves. *Scientific American* 199 (December): 105–112.

Higgins, John, and Douglas Campbell. 1994. Mathematical certificates. *Mathematics Magazine* 67 (February): 21–28.

Mahoney, Michael S. 1994. *The mathematical career of Pierre de Fermat*, 2nd ed. Princeton, N.J.: Princeton University Press.

Mollin, R. A. 1997. Prime-producing quadratics. *American Mathematical Monthly* (June/July): 529–544.

Peterson, Ivars. 1995. Quantum bits. *Science News* 147 (January 14): 30–31.

———. 1994. Major-league sieving for faster factoring. *Science News* 146 (July 30): 71.

———. 1994. Opening a quantum door on computing. *Science News* 145 (May 14): 308.

———. 1994. Team sieving cracks a huge number. *Science News* 145 (May 7): 292.

———. 1993. Dubner's primes. *Science News* 144 (November 20): 331.

———. 1992. Primality tests: An infinity of exceptions. *Science News* 142 (September 19): 182.

———. 1992. Striking pay dirt in prime-number terrain. *Science News* 141 (April 4): 213.

———. 1988. Cracking the 100-digit factoring barrier. *Science News* 134 (October 22): 163.

———. 1988. Priming for a lucky strike. *Science News* 133 (February 6): 85.

———. 1985. Prime time for supercomputers. *Science News* 128 (September 28): 199.

———. 1985. Uncommon factoring. Science News 127 (March 30): 202–203.

———. 1985. A curving path toward faster factoring. *Science News* 127 (March 9): 151.

———. 1984. Faster factoring for cracking computer security. *Science News* 125 (January 14): 20.

———. 1982. Quickening the pursuit of primes. *Science News* 121 (March 6): 158.

———. 1981. Whom do you trust? *Science News* 120 (September 26): 205–206.

Pomerance, Carl. 1997. A tale of two sieves. *Notices of the American Mathematical Society* 43 (December): 1473–1485.

———. 1981. Recent developments in primality testing. *Mathematical Intelligencer* 3, no. 3: 97–105.

Ribenboim, Paulo. 1996. *The new book of prime number records*. New York: Springer-Verlag.

———. 1995. Selling primes. *Mathematics Magazine* 68 (June): 175–182.

Schroeder, M. R. 1984. *Number theory in science and communication*. New York: Springer-Verlag.

Vanden Eynden, Charles. 1989. Flipping a coin over the telephone. *Mathematics Magazine* 62 (June): 167–172.

Williams, H. C. 1984. Factoring on a computer. *Mathematical Intelligencer* 6, no. 3: 29–36.

3 TWISTS OF SPACE

Adams, Colin C. 1994. *The knot book: An elementary introduction to the mathematical theory of knots*. New York: W. H. Freeman.

Almgren, Frederick J., Jr., and Jean E. Taylor. 1976. The geometry of soap films and soap bubbles. *Scientific American* 235 (July): 82–93.

Boys, C. V. 1959. *Soap bubbles: Their colors and the forces which mould them*. New York: Dover.

Callahan, Michael J., David Hoffman, and James T. Hoffman. 1988. Computer graphics tools for the study of minimal surfaces. *Communications of the* ACM 31 (June): 648–661.

Corrigan, Edward. 1993. Knot physics. *Physics World* 6 (June): 41–47.

Flam, Faye. 1989. Frothy physics. *Science News* 136 (July 29): 72–73, 76.

Garmon, L. 1982. Möbius molecule: Synthesis with a twist. *Science News* 122 (July 17): 36.

Hass, Joel, and Roger Schafly. 1996. Bubbles and double bubbles. *American Scientist* 84 (September-October): 462–467.

Hildebrandt, Stefan, and Anthony Tromba. 1996. *The parsimonious universe: Shape and form in the natural world*. New York: Copernicus.

Hoffman, David. 1986. The computer-aided discovery of new embedded minimal surfaces. *Mathematical Intelligencer* 9, no. 3: 8–21.

Jones, Vaughan F. R. 1990. Knot theory and statistical mechanics. *Scientific American* 263 (November): 98–103.

Kanigel, Robert. 1993. Bubble, bubble. *The Sciences* (May/June): 32–38.

Lickorish, W. B. R., and K. C. Millett. 1988. The new polynomial invariants of knots and links. *Mathematics Magazine* 61 (February): 3–23.

Morgan, Frank. 1996. What is a surface? *American Mathematical Monthly* 103 (May): 374–375.

_____. 1994. Mathematicians, including undergraduates, look at soap bubbles. *American Mathematical Monthly* 101 (April): 343–351.

_____. 1992. Minimal surfaces, crystals, shortest networks, and undergraduate research. *Mathematical Intelligencer* 14, no. 3: 37–44.

_____. 1986. Soap films and problems without unique solutions. *American Scientist* 74 (May-June): 232–235.

Neuwirth, Lee. 1979. The theory of knots. *Scientific American* 240 (June): 110–124.

Peterson, Ivars. 1996. Tying knots to tubular geometry, DNA loops. *Science News* 150 (November 16): 310.

_____. 1996. The song in the stone. *Science News* 149 (February 17): 110–111.

_____. 1995. Toil and trouble over double bubbles. *Science News* 148 (August 12): 101.

_____. 1994. Constructing a stingy scaffolding for foam. *Science News* 145 (March 25): 149.

_____. 1992. Putting a handle on minimal helicoid. *Science News* 142 (October 24): 276.

_____. 1992. Knotty views. *Science News* 141 (March 21): 186–187.

_____. 1989. Knot physics. *Science News* 135 (March 18): 174.

_____. 1989. Linking doughnuts, soda straws, and energy. *Science News* 135 (February 4): 70.

_____. 1988. Tying up a knotty loose end. *Science News* 134 (October 29): 283.

_____. 1988. The crystalline face of soap films. *Science News* 135 (August 27): 135.

_____. 1988. Unknotting a tangled tale. *Science News* 133 (May 21): 328–330.

_____. 1987. A metal's many faces. *Science News* 131 (January 31): 76–77.

_____. 1985. Untangling a knotty problem. *Science News* 128 (October 26): 266.

_____. 1985. Three bites in a doughnut. *Science News* 127 (March 16): 168–169.

Steen, Lynn Arthur. 1975. Solving the great bubble mystery. *Science News* 108 (September 20): 186–187.

Stewart, Doug. 1987. Skylines of fabric. *Technology Review* 90 (January): 60–67.

Sumners, De Witt. 1995. Lifting the curtain: Using topology to probe the hidden action of enzymes. *Notices of the American Mathematical Society* 42 (May): 528–537.

Taylor, J. E., and J. W. Cahn. 1986. Catalog of saddle shaped surfaces in crystals. *Acta Metallurgica* 34, no. 1: 1–12.

Watson, Andrew. 1991. Twists, tangles, and topology. *New Scientist* 132 (October 5): 42–46.

Wu, C. 1996. Chemicals get tied up in complex knots. *Science News* 150 (October 12): 231.

4 SHADOWS FROM HIGHER DIMENSIONS

Abbott, Edwin A. 1991. *Flatland: A romance of many dimensions*. Princeton, N.J.: Princeton University Press.

Asimov, Daniel. 1995. There's no space like home. *The Sciences* 35 (September/October): 20–25.

Banchoff, Thomas F. 1990. *Beyond the third dimension: Geometry, computer graphics, and higher dimensions*. New York: Scientific American Library.

Cole, K. C. 1993. Escape from 3-D. *Discover* 14 (July): 52–62.

Donaldson, S. K. 1996. The Seiberg-Witten equations and 4-manifold topology. *Bulletin of the American Mathematical Society* 33 (January): 45–73.

Francis, George K. 1987. *A topological picturebook*. New York: Springer-Verlag.

Gardner, Martin. 1992. Spheres and hyperspheres. In *Mathematical circus*. Washington, D. C.: Mathematical Association of America.

Henderson, Linda D. 1983. *The fourth dimension and non-Euclidean geometry in art*. Princeton, N. J.: Princeton University Press.

Heppenheimer, T. A. 1988. The mathematics of manifolds. *Mosaic* 19, no. 2: 32–43.

Koçak, Hüseyin, and David Laidlaw. 1987. Computer graphics and the geometry of S3. *Mathematical Intelligencer* 9, no. 1: 8–10.

Peterson, Ivars. 1989. A different dimension. *Science News* 135 (May 27): 328–330.

_____. 1984. Shadows from a higher dimension. *Science News* 126 (November 3): 284–285.

Smale, Steve. 1990. The story of the higher dimensional Poincaré conjecture (What actually happened on the beaches of Rio). *Mathematical Intelligencer* 12, no. 2: 44–51.

Steen, Lynn A. 1982. Twisting and turning in space. *Science News* 122 (July 17): 42–44.

Thurston, William P., and Jeffrey R. Weeks. 1984. The mathematics of three-dimensional manifolds. *Scientific American* 251 (July) 108–120.

Weeks, Jeffrey R. 1985. *The shape of space: How to visualize surfaces and three-dimensional manifolds*. New York: Marcel Dekker.

Witten, Edward. 1997. Duality, spacetime, and quantum mechanics. *Physics Today* 50 (May): 28–33.

5 ANTS IN LABYRINTHS

Barnsley, Michael F. 1996. Fractal image compression. *Notices of the American Mathematical Society* 43 (June): 657–662.

Barnsley, Michael F., V. Ervin, D. Hardin, and J. Lancaster. 1986. Solution of an inverse problem of fractals and other sets. *Proceedings of the National Academy of Sciences* (USA) 83 (April): 1975–1977.

Dewdney, A. K. 1986. Of fractal mountains, graftal plants, and other computer graphics at Pixar. *Scientific American* 255 (December): 14–20.

Gardner, Martin. 1989. Mandelbrot's fractals. In *Penrose tiles to trapdoor ciphers*. New York: W. H. Freeman.

Hannabuss, Keith. 1996. Forgotten fractals. *Mathematical Intelligencer* 18, no. 3: 28–31.

Mandelbrot, Benoit B. 1982. *The fractal geometry of nature*. New York: W. H. Freeman.

Peterson, Ivars. 1996. The shapes of cities. *Science News* 149 (January 6): 8–9.

———. 1993. From surface scum to fractal swirls. *Science News* 143 (January 23): 53.

———. 1987. Packing it in. *Science News* 131 (May 2): 283–285.

———. 1984. Ants in labyrinths and other fractal excursions. *Science News* 125 (January 21): 42–43.

Pickover, Clifford A., ed. 1996. *Fractal horizons: The future use of fractals*. New York: St. Martin's Press.

Raloff, Janet. 1982. Computing for art's sake. *Science News* 122 (November 20): 328–331.

Rucker, Rudy. 1987. *Mind tools*. New York: Houghton Mifflin.

Steen, Lynn Arthur. 1977. Fractals: A world of nonintegral dimensions. *Science News* 131 (March 21): 184.

Stoppard, Tom. 1993. *Arcadia*. London: Faber and Faber.

Thomsen, Dietrick E. 1987. Fractals: Magical fun or revolutionary science? *Science News* 131 (March 21): 184.

———. 1982. A place in the sun for fractals. *Science News* 121 (January 9): 28–30.

Weisburd, S. 1985. Fractals, fractures, and faults. *Science News* 127 (May 4): 279.

6 THE DRAGONS OF CHAOS

Allgood, Kathleen T., Tim D. Sauer, and James A. Yorke. 1996. *Chaos: An introduction to dynamical systems*. New York: Springer-Verlag.

Baker, Gregory L., and Jerry P. Gollub. 1990. *Chaotic dynamics: An introduction*. Cambridge, England: Cambridge University Press.

Crutchfield, James P., J. Doyne Farmer, Norman H. Packard, and Robert S. Shaw. 1986. Chaos. *Scientific American* 255 (December): 46–57.

Devaney, Robert L. 1987. Chaotic bursts in nonlinear dynamical systems. *Science* 235 (January 16): 342–345.

Ditto, William L., and Louis M. Pecora. 1993. Mastering chaos. *Scientific American* 269 (August): 78–84.

Ford, Joseph. 1983. How random is a coin toss? *Physics Today* 36 (April): 40–47.

Glanz, James. 1997. Mastering the nonlinear brain. *Science* 277 (September 19): 1758–1760.

Gleick, James. 1987. *Chaos: The making of a new science*. New York: Viking.

Gulick, Denny. 1992. *Encounters with chaos*. New York: McGraw-Hill.

Hofstadter, Douglas R. 1981. Strange attractors: Mathematical patterns delicately poised between order and chaos. *Scientific American* 245 (November): 22–43.

Hubbard, John H. 1986. Order in chaos. *Engineering: Cornell Quarterly* 20, no. 3: 20–26.

Jensen, Roderick V. 1987. Classical chaos. *American Scientist* 75 (March-April): 168–181.

Kadanoff, Leo P. 1983. Roads to chaos. *Physics Today* 36 (December): 46–53.

Lorenz, Edward N. 1993. *The essence of chaos*. Seattle: University of Washington Press.

Ott, Edward, and Mark Spano. 1995. Controlling chaos. *Physics Today* 48 (May): 34–40.

Ott, Edward, Celso Grebogi, and James A. Yorke. 1990. Controlling chaos. *Physical Review Letters* 64 (March 12): 1196–1199.

Peitgen, H.-O., and P. H. Richter. 1986. *The beauty of fractals*. New York: Springer-Verlag.

Pendick, Daniel. 1993. Chaos of the mind. *Science News* 143 (February 27): 138–139.

Peterson, Ivars. 1993. *Newton's clock: Chaos in the solar system*. New York: W. H. Freeman.

_____. 1992. Chaos in the clockwork. *Science News* 141 (February 22): 120–121.

_____. 1991. Bordering on infinity. *Science News* 140 (November 23): 331.

_____. 1991. Ribbon of chaos. *Science News* 139 (January 26): 60–61.

_____. 1988. In the shadows of chaos. *Science News* 134 (December 3): 360.

_____. 1987. Portraits of equations. *Science News* 132 (September 19): 184–185.

_____. 1987. Zeroing in on chaos. *Science News* 131 (February 28): 137–139.

_____. 1985. The chaos of war. *Science News* 127 (January 5): 13.

_____. 1984. Escape into chaos. *Science News* 125 (May 26): 328.

_____. 1983. Pathways to chaos. *Science News* 124 (July 30): 76–77.

Ruelle, David. 1991. *Chance and chaos.* Princeton, N.J.: Princeton University Press.
_____. Strange attractors. *Mathematical Intelligencer* 2, no. 3: 126–137.

7 LIFE STORIES

Berlekamp, Elwyn R., John H. Conway, and Richard K. Guy. 1982. *Winning ways for your mathematical plays.* San Diego: Academic Press.

Casti, John L. 1996. *Would-be worlds: How simulation is changing the frontiers of science.* New York: Wiley.

Dewdney, A. K. 1989. Two dimensional Turing machines and tur-mites make tracks on a plane. *Scientific American* 261 (September): 180–183.

_____. Building computers in one dimension sheds light on irreducibly complicated phenomena. *Scientific American* 252 (May): 18–30.

Durrett, Rick. 1991. Some new games for your computer. *Nonlinear Science Today* 1, no. 4: 1–7.

Epstein, Joshua M., and Robert Axtell. 1996. *Growing artificial societies: Social science from the bottom up.* Washington, D. C.: Brookings Institution Press.

Gale, David. 1993. The industrious ant. *Mathematical Intelligencer* 15, no. 2: 54–55.

Gale, David, Jim Propp, Scott Sutherland, and Serge Troubetzkoy. 1995. Further travels with my ant. *Mathematical Intelligencer* 17, no. 3: 48–56.

Gardner, Martin. 1983. *Wheels, life, and other mathematical amusements.* New York: W. H. Freeman.

Hotz, Robert Lee. 1997. A study in complexity. *Technology Review* 100 (October): 22–29.

Peterson, Ivars. 1996. The gods of Sugarscape. *Science News* 150 (November 23): 332–333.

_____. 1995. Travels of an ant. *Science News* 148 (October 28): 280–281, 287.

_____. 1987. Forest fires, barnacles, and trickling oil. *Science News* 132 (October 3): 220–223.

Poundstone, William. 1985. *The recursive universe.* New York: William Morrow.

Propp, Jim. 1994. Further ant-ics. *Mathematical Intelligencer* 16, no. 1: 37–42.

Stewart, Ian. 1994. The ultimate in anty-particles. *Scientific American* 271 (July): 104–107.

Wolfram, Stephen. 1984. Cellular automata as models of complexity. *Nature* 311 (October 4): 419–424.

_____. 1984. Computer software in science and mathematics. *Scientific American* 251 (September): 188–203.

8 IN ABSTRACT TERRAIN

Aczel, Amir D. 1996. *Fermat's last theorem: Unlocking the secret of an ancient mathematical problem.* New York: Four Walls.

Blum, Manuel, Alfredo de Santis, Silvio Micali, and Giuseppe Persiano. 1991. Noninteractive zero-knowledge. *SIAM Journal of Computation* 20 (December): 1084–1118.

Further Reading

Cipra, Barry. 1992. "Transparent" proofs help solve opaque problems. *Science* 256 (May 15): 970–971.

Dewdney, A. K. 1984. A computer trap for the busy beaver, the hardest-working Turing machine. *Scientific American* 251 (August): 19–23.

Epstein, David, and Silvio Levy. 1995. Experimentation and proof in mathematics. *Notices of the American Mathematical Society* 42 (June): 670–674.

Fagin, Ronald, Moni Naor, and Peter Winkler. 1996. Comparing information without leaking it. *Communications of the* ACM 39 (May): 77–85.

Graham, R. L., and J. Nesetril. 1996. *The mathematics of Paul Erdös*, vol. 1. New York: Springer-Verlag.

Hopcroft, John E. 1984. Turing machines. *Scientific American* 250 (May): 86–98.

Kolata, Gina. 1986. Math proof refuted during Berkeley scrutiny. *Science* 234 (December 19): 1498–1499.

Landau, Susan. 1988. Zero knowledge and the Department of Defense. *Notices of the American Mathematical Society* 35 (January): 5–12.

McGeoch, Catherine C. 1993. Zero-knowledge proofs. *American Mathematical Monthly* 100 (August-September): 682–685.

Peterson, Ivars. 1997. Computers and proof. *Science News* 151 (March 22): 176–177.

_____. 1994. Fermat's famous theorem: Proved at last? *Science News* 146 (November 5): 295.

_____. 1994. Last word not yet in on Fermat's conjecture. *Science News* 145 (June 25): 406.

_____. 1993. Fermat proof flaw: Fixing the details. *Science News* 144 (December 18/25): 406.

_____. 1993. A curvy path leads to Fermat's last theorem. *Science News* 144 (July 3): 5–6.

_____. 1992. Holographic proofs. *Science News* 141 (June 6): 382–383.

_____. 1990. Beyond chaos: Ultimate predictability. *Science News* 137 (May 26): 327.

_____. 1989. The business of busy beavers. *Science News* 136 (September 16): 191.

_____. 1988. Doubts about Fermat solution. *Science News* 133 (April 9): 230.

_____. 1988. Fermat's last theorem: A promising approach. *Science News* 133 (March 19): 180–181.

_____. 1988. Computing a bit of security. *Science News* (January 16): 38.

_____. 1987. Closing in on Fermat's last theorem. *Science News* 131 (June 20): 397.

_____. 1986. Keeping secrets. *Science News* 130 (August 30): 140.

_____. 1986. Waiting for the Poincaré proof. *Science News* 129 (April 5): 215.

_____. 1985. Looking for the busy beaver. *Science News* 127 (February 9): 89.

Rourke, Colin, and Ian Stewart. 1986. Poincaré's perplexing problem. *New Scientist* 111 (September 4): 41–45.

Sackett, P. D. 1983. New proof handles old math problem. *Science News* 124 (July 23): 58.

Sangalli, Arturo. 1993. The easy way to check hard maths. *New Scientist* 138 (May 8): 24–28.

Further Reading

Singh, Simon. 1997. *Fermat's enigma: The quest to solve the world's greatest mathematical problem*. New York: Walker.

Stillwell, John. 1995. Elliptic curves. *American Mathematical Monthly* 104 (November): 831–837.

Taubes, Gary. 1987. What happens when hubris meets nemesis. *Discover* 8 (July): 66–77.

Thurston, William P. 1994. On proof and progress in mathematics. *Bulletin of the American Mathematical Society* 30 (April): 161–177.

— SOURCES OF ILLUSTRATIONS —

Figure **1.1** (*bottom*) Created by Edward F. Moore, University of Wisconsin (*Scientific American*, October 1977, page 109)

Figure **1.3** Charles S. Peskin, New York University

Figure **1.4** Charles S. Peskin and David M. McQueen, New York University

Figure **1.7** *Scientific American*, January 1977, page 115

Figure **1.8** *The Mathematical Intelligencer,* © 1979 by Springer-Verlag

Figure **1.9** *Scientific American*, January 1977, page 116

Figure **1.11** Seymour Haber, National Institute of Standards and Technology

Figure **1.13** Michel Duneau and André Katz, École Polytechnique, Palaiseau, France

Figure **1.14** Paul J. Steinhardt, University of Pennsylvania

Figure **2.1** *Scientific American*, March 1964, page 122

Figure **2.4** *Scientific American*, March 1964, page 120

Figure **3.1** Frei Otto, Institut für leichte Flächentragwerke, Stuttgart

Figure **3.3** Stefan Hildebrandt and Anthony Tromba, *Mathematics and Optimal Form,* © 1985 by Scientific American Books

Figure **3.4** Stefan Hildebrandt and Anthony Tromba, *Mathematics and Optimal Form,* © 1985 by Scientific American Books

Figure **3.11** James T. Hoffman and David Hoffman

Figure **3.12** James T. Hoffman and David Hoffman

Figure **3.13** David Hoffman, Hermann Karcher, and Fusheng Wei. Graphics from James T. Hoffman, Mathematical Sciences Research Institute

Figure **3.15** John M. Sullivan

Figure **3.16** *Left*: Adapted from photo courtesy of R. J. Gray, Oak Ridge National Laboratory *Right*: Cyril Stanley Smith, *A Search for Structure*, MIT Press, © 1981 by the Massachusetts Institute of Technology

Figure **3.17** Jean E. Taylor, Rutgers University

Figure **3.18** Jean E. Taylor, Rutgers University

Sources of Illustrations

Figure 3.19 *Scientific American*, June 1979, page 112

Figure 3.20 Joan S. Birman, Columbia University

Figure 3.24 Nicholas R. Cozzarelli, University of California at Berkeley, and Steven A. Wasserman, University of Texas at Austin

Figure 3.25 Nicholas R. Cozzarelli, University of California at Berkeley, and Steven A. Wasserman, University of Texas at Austin

Figure 4.3 *Scientific American*, June 1987, page 112

Figure 4.4 *Scientific American*, April 1986, page 20

Figure 4.6 Thomas F. Banchoff, Brown University

Figure 4.7 Lynn Steen, ed., *Mathematics Today*, Springer-Verlag, © 1978 by the Conference Board of the Mathematical Sciences

Figure 4.8 *Scientific American*, April 1986, page 20

Figure 4.9 Thomas F. Banchoff, Hüseyin Koçak, and David Laidlaw, Brown University

Figure 4.10 David Laidlaw and Hüseyin Koçak, Brown University

Figure 4.12 First appeared in *New Scientist*, London, the weekly review of science and technology

Figure 5.1 Benoit B. Mandelbrot, *The Fractal Geometry of Nature*, W. H. Freeman and Company, © 1982 by Benoit B. Mandelbrot

Figure 5.2 Benoit B. Mandelbrot, *The Fractal Geometry of Nature*, W. H. Freeman and Company, © 1982 by Benoit B. Mandelbrot

Figure 5.3 Benoit B. Mandelbrot, *The Fractal Geometry of Nature*, W. H. Freeman and Company, © 1982 by Benoit B. Mandelbrot

Figure 5.4 Leonard M. Blumenthal and Karl Menger, *Studies in Geometry*, © 1970 by W. H. Freeman and Company

Figure 5.5 Alvy Ray Smith, Lucasfilm Ltd. (*Scientific American*, September 1984, page 156)

Figure 5.7 Benoit B. Mandelbrot, *The Fractal Geometry of Nature*, W. H. Freeman and Company, © 1982 by Benoit B. Mandelbrot

Figure 5.8 Peter Oppenheimer, New York Institute of Technology

Figure 5.10 Michael F. Barnsley

Figure 5.12 Michael F. Barnsley

Sources of Illustrations

Figure 5.13 Benoit B. Mandelbrot, *The Fractal Geometry of Nature*, W. H. Freeman and Company, © 1982 by Benoit B. Mandelbrot

Figure 5.14 Benoit B. Mandelbrot, *The Fractal Geometry of Nature*, W. H. Freeman and Company, © 1982 by Benoit B. Mandelbrot

Figure 5.15 Paul Meakin, Du Pont Experimental Station

Figure 5.16 Paul Meakin, Du Pont Experimental Station

Figure 5.17 *Top*: Roy Richter, GM Research Laboratories *Bottom*: Nancy Hecker and David G. Grier, University of Michigan (*Scientific American*, January 1987, page 98)

Figure 6.2 Edward N. Lorenz, Massachusetts Institute of Technology

Figure 6.3 *Scientific American*, July 1987, page 110

Figure 6.4 John Guckenheimer, Cornell University

Figure 6.5 Benoit B. Mandelbrot, *The Fractal Geometry of Nature*, W. H. Freeman and Company, © 1982 by Benoit B. Mandelbrot

Figure 6.6 Reprinted with permission from *The Beauty of Fractals* by Heinz-Otto Peitgen and Peter H. Richter, Springer-Verlag, 1986

Figure 6.7 Benoit B. Mandelbrot, *The Fractal Geometry of Nature*, W. H. Freeman and Company, © 1982 by Benoit B. Mandelbrot

Figure 6.8 Reprinted with permission from *The Beauty of Fractals* by Heinz-Otto Peitgen and Peter H. Richter, Springer-Verlag, 1986

Figure 6.9 Reprinted with permission from *The Beauty of Fractals* by Heinz-Otto Peitgen and Peter H. Richter, Springer-Verlag, 1986

Figure 6.12 Reprinted with permission from *The Beauty of Fractals* by Heinz-Otto Peitgen and Peter H. Richter, Springer-Verlag, 1986

Figure 6.13 S. Burns, H. Benzinger, and J. Palmore, University of Illinois

Figure 6.14 Robert L. Devaney, Boston University

Figure 7.2 *Scientific American*, November 1984, page 40

Figure 7.3 *Abacus*, © 1987 by Springer-Verlag

Figure 7.5 Stephen Wolfram

Figure 7.6 Stephen Wolfram

Figure 7.7 Stephen Wolfram

Figure 7.8 Stephen Wolfram

Sources of Illustrations

Figure 7.9 Stephen Wolfram

Figure 7.10 Rick Durrett, Cornell University

Figure 7.12 Rick Durrett, Cornell University

Figure 7.13 Rick Durrett, Cornell University

Figure 7.14 Scott Sutherland, State University of New York at Stony Brook

Figure 7.15 Scott Sutherland, State University of New York at Stony Brook

Figure 7.16 Scott Sutherland, State University of New York at Stony Brook

Figure 7.17 Joshua M. Epstein and Robert Axtell, *Growing Artificial Societies: Social Science from the Bottom Up*, Brookings Institution, 1996

Figure 8.4 *Scientific American*, August 1984, page 20

Plate 1 Paul J. Steinhardt, University of Pennsylvania

Plate 2 Paul J. Steinhardt, University of Pennsylvania

Plate 3 Minimal Möbius band discovered by W. H. Meeks. Computer-generated image © 1987 by James T. Hoffman and David Hoffman

Plate 4 Genus-1, Costa-Hoffman-Meeks embedded minimal surface. Computed-generated image © 1987 by James T. Hoffman and David Hoffman

Plate 5 The core of the four-lobed Wente torus. Compiled by J. Spruck, A. Eydeland, and M. Callahan; image © 1987 by James T. Hoffman

Plate 6 Produced at Brown University by Hüseyin Koçak and David Laidlaw in collaboration with T. Banchoff, F. Bisshopp, and D. Margolis

Plate 7 Richard F. Voss, IBM Research

Plate 8 Paul Meakin, Du Pont Experimental Station

Plate 9 James P. Crutchfield, University of California at Berkeley (*Scientific American*, December 1986, page 57)

Plate 10 Generated on NASA's Massively Parallel Processor by Edward Seller, NASA Goddard Space Flight Center

Plate 11 S. Burns, H. Benzinger, and J. Palmore, University of Illinois

Plate 12 S. Burns, H. Benzinger, and J. Palmore, University of Illinois

Plate 13 Robert L. Devaney, Boston University

Plate 14 Stephen Wolfram (*Scientific American*, September 1984, page 199)

Plate 15 Norman H. Packard (*Scientific American*, September 1984, page 189)

Plate 16 Scott Sutherland, State University of New York at Stony Brook

INDEX

Index

Index

Index

Index

Index

Index